1*a* **Mid-summer flowering of** *Celmisia–Poa*
tall alpine herbfield near Lake Cootapatamba

CSIRO / COLLINS AUSTRALIA

A B Costin
M Gray
C J Totterdell
D J Wimbush

Kosciusko Alpine Flora

© CSIRO 1979 Reprinted 1980   Published jointly by
CSIRO, 314 Albert Street, East Melbourne, Australia 3002 and
William Collins Pty Ltd, 55 Clarence Street, Sydney, Australia 2000

The prefatory quotation by Sir Peter Medawar is from 'Induction and
Intuition in Scientific Thought' published by Methuen and Co Ltd
Text set in Bembo by the Dova Type Shop, Melbourne, Victoria
Printed and bound at Griffin Press Ltd, Adelaide, South Australia
CSIRO Editorial and Production: Basil Walby, Margaret Walkom,
Barbara Williams   Designed by Alison Forbes

Kosciusko alpine flora

Index
Simultaneously published, Sydney: Collins
Bibliography
ISBN 0 643 02473 5   ISBN 0 643 02474 3 Paperback

1. Plant communities — New South Wales — Mount Kosciusko
district. 2. Alpine flora — New South Wales — Mount Kosciusko
district — Identification. 3. Botany — New South Wales — Mount
Kosciusko district. I. Costin, Alec Baillie. II. Commonwealth
Scientific and Industrial Research Organization.

581.5'0994'4 [7]

Library of Congress Card No. 80-670062

To Margaret, Betty, Val and Robyn, who have helped
us overcome the many difficulties in producing this book

Contents

Foreword

Kosciusko is a region of great scientific and educational interest. Up to now, a comprehensive account of the Kosciusko environment and its alpine plants has not been available. This book draws on well over a century of research, from the botanical explorations of the 1850s up to the present day. It should be a source of information to all people interested in high mountain environments and the plants which grow there. It will be useful to those who want to know only a little more about the flowers they are seeing, as well as to those with a deeper interest in the plants and their environments. The illustrations will delight all who love mountain scenery and enjoy mountain flowers.

Kosciusko Alpine Flora is a fitting commemoration of the 50th anniversary of the Division of Plant Industry, one of CSIRO's longest established research groups.

W. J. PEACOCK   *Chief, Division of Plant Industry, CSIRO, Canberra*

The purpose of scientific enquiry is not to compile an inventory of factual information, nor to build up a totalitarian world picture of natural Laws in which every event that is not compulsory is forbidden. We should think of it rather as a logically articulated structure of justifiable beliefs about nature. It begins as a story about a Possible World — a story which we invent and criticize and modify as we go along, so that it ends by being, as nearly as we can make it, a story about real life.

P. B. MEDAWAR

Preface

Kosciusko Alpine Flora has been a combined effort in which we have endeavoured to bring together the results of our individual knowledge, insights and experience of the Kosciusko environment and its plant life. All of us are impressed by the outstanding beauty and scientific interest of Australia's geographically restricted alpine flora, of which Kosciusko possesses the finest example, and we are dedicated to its conservation. In sharing the results of our enthusiasm and experience with others – and by trying to present the beauty as well as the science – we trust that they, too, will learn to understand and appreciate this unique and wonderful flora and, in doing so, become committed to conserving it. In producing this book, we have aimed to combine scientific merit and popular appeal. We believe that such a combination is possible and that science should keep the community as well as the scientist in mind.

Kosciusko Alpine Flora contains an introductory section on the general features of the area and its geologic and climatic history with special reference to the flora. This is followed by sections on human history and use and the general characteristics of the alpine flora, its probable origin, and its relation to other alpine floras. Descriptions and photographs are provided for each main plant community of which the individual species of the flora are members and to which they are subsequently referred. This information should assist readers to understand the ecological relationships of each species of plant as well as to locate it in the field. Keys are provided for the identification of the species. Each species is then described with reference to colour plates.

Large-scale colour maps of the Kosciusko area delineate the distribution of the main vegetation communities together with important physiographic and other habitat features of the area such as lakes, cliffs, screes, tracks and eroded areas. The maps were compiled from aerial vertical black-and-white photography supplemented by specially flown oblique colour photography. The glossary, index and extensive bibliography will enable readers to follow up particular aspects of the book in greater detail.

Alec Costin wrote the general text and he and Dane Wimbush supplied the ecological data. Max Gray is responsible for the taxonomic part. With the exception of the historical photographs, and the colour plates 9 and 31 taken by Alec Costin, the photography is by Colin Totterdell, who also contributed to both the taxonomic and ecological aspects through extensive field collecting and observations. Dane Wimbush contributed the colour vegetation maps.

Numerous fellow botanists and ecologists have provided specimens and information on the plant species. The Director and the staff of the New South Wales National Herbarium have been characteristically most helpful and cooperative. Dr Elizabeth Edgar and Dr V. D. Zotov, of the New Zealand Department of Scientific and Industrial Research, and Dr Winifred M. Curtis, of Tasmania, have been most generous in making available their specialized knowledge of various groups and in providing specimens for comparison. Responsibility for the use of the information rests with the authors and any errors which may have been included are theirs. The field work was assisted by the generosity of the officers of the New South Wales Soil Conservation Service, particularly Mr Roger Good (now with the National Parks and Wildlife Service of New South Wales), who arranged transport and accommodation for us at the Carruthers Hut on a number of occasions. We are also indebted to Dr Donald McVean, at the time a Senior Research Fellow at the Australian National University, Professor Kay Beamish, University of British Columbia, and Mrs Margaret Parris of the Monaro Conservation Society who led us to some of the rarer plants. Our special thanks go to Mr Gratton Wilson, Secretary of CSIRO, who first proposed the publication of the work by the Organization, and to Drs Lloyd Evans and Jim Peacock, successive Chiefs of the Division of Plant Industry, for their help and encouragement. Mrs Vi Taylor and her staff in the CSIRO Division of Plant Industry typing pool were always more than helpful in the formidable task of re-typing material as it was revised. We are most grateful for the help offered by Emeritus Professor A. D. Hope, Professors Frank Fenner, Lindsay Pryor and Donald Walker, all of the Australian National University; Professor Walker also provided facilities in the Department of Biogeography and Geomorphology during the final stages of preparing and collating the text. We are grateful for the valuable contributions made to the book by colleagues in the Illustration and Photography Sections of the Division of Plant Industry and thank Mr Dick Baas Becking, Mrs Sandie McIntosh and Lea O'Brien for preparing the maps and figures, and Mr Tibor Binder, Mrs Joan Simpson and Mr Emile Brunoro for their advice during the final selection of photographs.

Alec Costin   Max Gray   Colin Totterdell   Dane Wimbush

The Kosciusko Environment

The term 'alpine' refers to those areas and their components between the climatic limit of tree vegetation and the nival zone (i.e. the zone of permanent snow and ice cover). Some workers have questioned whether the Kosciusko area is truly alpine, in the sense that it occurs above the climatic treeline, or whether the treeline is determined by local, non-climatic conditions such as shallow or wet soils. They point to the occurrence of treelines at much higher elevations in parts of the northern hemisphere at similar latitudes to Kosciusko: for example, the treeline in the Colorado Rockies is at about 3000 metres, compared with only 1800 metres at Kosciusko.

This apparent discrepancy – and there are many others when treelines at similar latitudes in different parts of the world are compared – prompts a comparison of the various environmental features of different treeline localities, to see if there is a common denominator. Depth and duration of snow cover, lowest temperatures experienced, wind regime, soil conditions and many others have been compared and are found to vary greatly. But whether the treelines occur in the northern or southern hemisphere, in high mountains or in the subantarctic or arctic, the mean temperature of the warmest month is always remarkably similar at about 10°C; and this is also the case at Kosciusko (Costin 1968; Daubenmire 1954).

With mean mid-summer temperatures of about 10°C the physiological limits to tree growth are apparently reached. Here it is thought that 'solar energy is adequate only to meet the annual requirements for respiration plus the requirements for foliage renewal, with the result that none is left to permit the development and maintenance of a large mass of non-productive cells, as comprise the stem and root system of a normal tree' (Daubenmire 1954). Thus it is found that mountains with relatively oceanic (i.e. cool and moist) summers have lower treelines than more continental mountains with greater extremes of temperature. All of the Australasian treelines, including those at Kosciusko, are not far from the sea which has a moderating effect on temperature in summer as well as winter. Furthermore, when treelines from the Australasian region are considered together (Costin 1968) they are found to increase in elevation by about 110 m for each degree decrease in latitude, which is almost precisely the pattern in other regions of the world. Thus, by world standards, the Kosciusko environment above the treeline at about 1830 m up to the summit at 2228 m is truly alpine in character. Although restricted in area in Australia, alpine environments cover a large part of the Earth's surface, and have been even more extensive in the past; Osburn and Wright (1967) and Ives and Barry (1974) collate some of the information available for other countries.

Fig. 1 shows the distribution of the so-called 'snow country' of Australia. It includes both the more restricted alpine areas above treeline and the subalpine areas down to about 300–500 m below treeline. This

Fig. 1 Map showing Kosciusko in relation to other subalpine and alpine areas of mainland Australia and Tasmania. Although only the flora of the Kosciusko alpine area is featured in this book, most of the species also occur elsewhere in the high country.

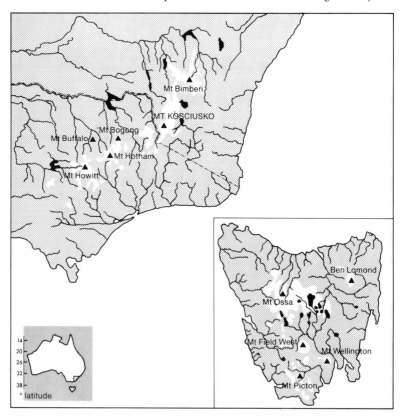

Vegetation and associated features of the Kosciusko Alpine area (after Wimbush and Costin 1973). Altitude: 1525–2135 metres

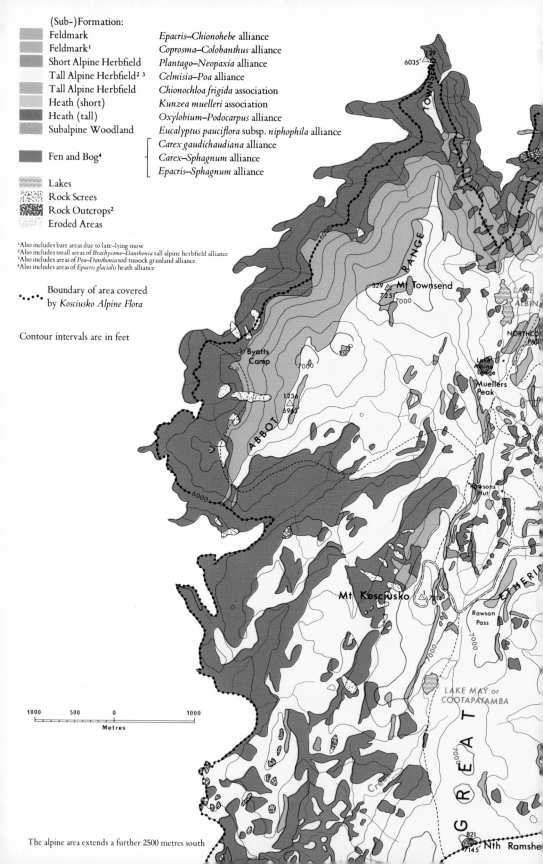

(Sub-)Formation:

Feldmark	*Epacris–Chionohebe* alliance
Feldmark[1]	*Coprosma–Colobanthus* alliance
Short Alpine Herbfield	*Plantago–Neopaxia* alliance
Tall Alpine Herbfield[2,3]	*Celmisia–Poa* alliance
Tall Alpine Herbfield	*Chionochloa frigida* association
Heath (short)	*Kunzea muelleri* association
Heath (tall)	*Oxylobium–Podocarpus* alliance
Subalpine Woodland	*Eucalyptus pauciflora* subsp. *niphophila* alliance

Fen and Bog[4]

Carex gaudichaudiana alliance
Carex–Sphagnum alliance
Epacris–Sphagnum alliance

Lakes
Rock Screes
Rock Outcrops[2]
Eroded Areas

[1] Also includes bare areas due to late-lying snow
[2] Also includes small areas of *Brachycome–Danthonia* tall alpine herbfield alliance
[3] Also includes areas of *Poa–Danthonia* sod tussock grassland alliance
[4] Also includes areas of *Epacris glacialis* heath alliance

· · · · · Boundary of area covered
by *Kosciusko Alpine Flora*

Contour intervals are in feet

The alpine area extends a further 2500 metres south

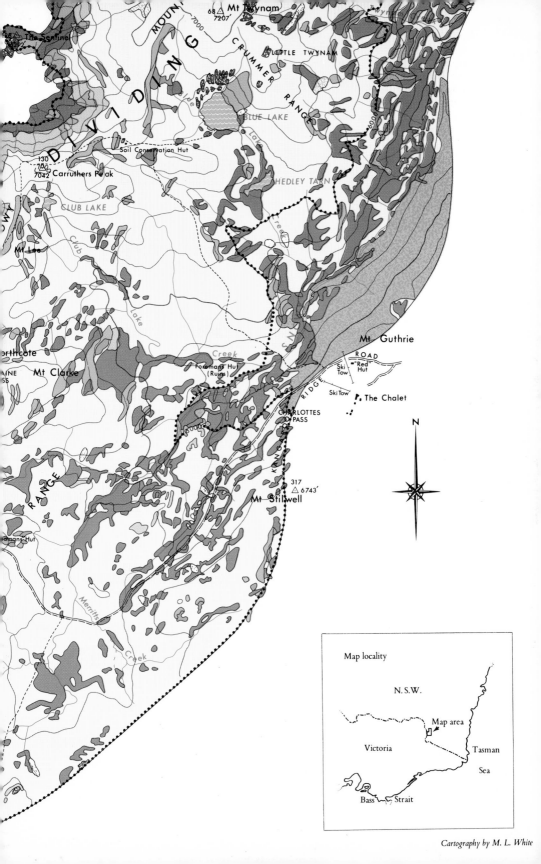

Cartography by M. L. White

The geographically restricted Kosciusko alpine area can be distinguished on this composite colour image of approximately 19 500 sq km of south-eastern Australia, obtained from the NASA LANDSAT I satellite (formerly called Earth Resources Technology Satellite I), by the characteristic north–south oriented semi-permanent snow patches along the lee slopes of the ranges. The straight valley of the Thredbo River can be seen on the right of the Kosciusko area, and the general relationship of the region to other major features such as the extensive montane and subalpine areas, Lakes Eucumbene and Jindabyne and the Snowy River can be seen. The relative smallness and isolation of the unique Kosciusko alpine environment is a significant feature of this remarkable picture. (The unusual red coloration, depicting vegetation, is typical of infrared-recording imagery.)

snow country occupies about 6500 sq km in Tasmania and 5200 on the
mainland, the combined area being only 0.15% of Australia as a whole.
The Snowy Mountains region, with about 2500 sq km of snow country,
is one of the most extensive of these areas. Of this, some 250 sq km are
truly alpine, including the main alpine area of about 100 sq km in the
Kosciusko Primitive Area, which is the region we are considering. Despite
its small size in relation to Australia as a whole (about 0.001%), the Kosciusko alpine area supports a rich, diverse and in part distinctive flora which
compares in beauty and interest with better-known alpine floras such
as those of the Rocky Mountains and the Swiss Alps. Some of the Kosciusko
alpines as shown in Table 3 (p. 47) are endemic (i.e. they occur nowhere
else), but many of them are also found throughout the snow country,
as defined in Fig. 1.

EVOLUTION   The Kosciusko area has not always been as it is today. There
has been a long history of change extending over hundreds of millions
of years. Some 450 million years ago Kosciusko was covered by a large
sea which extended over most of what is now eastern Australia. During
this period (the Ordovician), extensive sediments were deposited. Rem-
nants of these sediments (now much altered) are still to be seen as the
slates, phyllites, quartzites and schists forming part of the Kosciusko area
between Rawson Pass and Watson's Crags. Periods of folding, uplift and
sedimentation continued for millions of years during the Ordovician into
the Silurian and early Devonian periods when intrusion of granites and
folding and uplift of the area above sea level occurred. The oldest of
these granites, the most abundant rocks in the Kosciusko alpine area,
are about 390 million years old. Many millions of years of relative stability
of the Earth's crust followed, during which the uplifted areas were weath-
ered and slowly worn down to a fairly even peneplain surface with only
a few of the most resistant parts, including some of the Kosciusko peaks,
remaining above the general level. This long period of crustal stability
and erosion extended through most of the Carboniferous, Permian, Trias-
sic, Jurassic and Cretaceous periods until the beginning of the Tertiary
period about 60 million years ago.

The Tertiary period commenced an era of major uplift of eastern Australia during which the Kosciusko alpine area reached approximately its
present elevation. The uplifting, which continued spasmodically until
several million and perhaps as recently as about one million years ago,
also caused extensive fracturing and faulting of the rocks and gave the
rivers new erosive power. The major fracture patterns in the rocks provided zones of weakness along which many of the streams were able
to cut down more rapidly to establish the stream pattern of long straight

parallel courses which we see in the Kosciusko area today: the upper Snowy, Crackenback, Guthega and Munyang Rivers provide good examples. From fossil evidence we know that the climate during much of the Tertiary was warmer and wetter than it is now.

The Pleistocene period, probably commencing about two million years ago, was a time of generally colder climates throughout the world. At higher latitudes and altitudes glacial conditions developed, interspersed with warmer interglacials when the snow and ice cover largely or completely disappeared again. The Kosciusko area was likewise glaciated, although apparently weakly (Galloway 1963). The cirques, lakes, moraines, erratics and polished pavements seen at Kosciusko are products of these glacial conditions. Where the ice cover was thin or absent, the low temperatures also produced shattering of rocks, differential freezing and thawing of the soil with movement and accumulation of debris downslope, and other so-called periglacial effects both within and below the glaciated area itself (e.g. Costin and Polach 1971). For the last few thousand years (in the so-called Recent period) Kosciusko has been virtually ice-free, although some of the late-lasting snow patches sometimes persist for more than a year at a time.

Because the Pleistocene period was such an important influence in most parts of the world in the evolution of the present landscape, soils, flora and vegetation, and the fauna including man himself, it has received considerable study. Shortly, we shall attempt to review similar studies for Kosciusko. But first we should look briefly at the relation of the Kosciusko area and Australia as a whole to other land masses during the millions of years we have briefly reviewed because these relationships have been important in the origin of parts of the present-day alpine flora.

For more than a century biologists, commencing with the famous botanist Joseph Hooker (1860), have been intrigued by the similarity between the floras and faunas of certain parts of the world which are now widely separated (e.g. South America, New Zealand, the sub-Antarctic islands, Tasmania and the Australian Alps). These similarities go much further than similarities between individual species; sometimes whole groups of related species are involved. Such distributions are generally interpreted as indicating former contiguity of the separated land masses rather than chance distribution by long-distance dispersal as by birds, wind and water. However, until recently geologists believed that the former contiguities of some of these land masses, if in fact they ever occurred, must have been before the evolution of the flowering plants (which seem to have originated near the beginning of the Cretaceous period, about 100 million years ago). Thus for many years there was a stalemate between geological and biological opinion.

The Kosciusko alpine plateau is a much modified ancient peneplain surface, part of which is seen in this aerial view of the region looking south-east across Lake Albina, Mueller's Peak and Mt Kosciusko towards the forested mountains of the Upper Murray catchment. The valley of Lake Albina has been partially formed by glacial action and the lake itself is flanked and dammed by hummocky moraine. Mt Northcote is on the left of the lake and the rocky slopes of Mt Townsend are on the extreme right. Beyond Mt Northcote are Rawson's Valley, Rawson's Pass, Lake Cootapatamba and, to the left, Etheridge and Ramshead Ranges. Mt Kosciusko, the western slopes streaked with periglacial blockstreams, is in the centre of the picture with Wilkinson's Valley to the right. The apparently bare patches along the leeward east-facing slopes are feldmarks of the *Coprosma–Colobanthus* alliance and on the ridges, visible in the left foreground, is *Epacris–Chionohebe* feldmark.

Club Lake, with its associated cirque below Carruthers' Peak, is the smallest of the glacial lakes in the Kosciusko alpine area. The mouth of the lake is flanked by distinct moraine ridges. The Grey Mare Range is in the background behind Sentinel Peak and Watson's Crags.

But modern evidence is closing the gap (e.g. Jones 1971; Falvey 1972). Studies of the shape of the edges of continents and continental shelves and of the rocks which form them indicate that many areas now widely separated at one time fitted together. Furthermore, studies of the magnetism of the rocks concerned show that the position of some of the land masses has changed, and absolute dating of the rocks based on their residual radioactivity allows definite ages to be put on the times of former contiguity and subsequent drifting apart. Such evidence indicates that the southern land masses (including what is now Antarctica) were joined or were much closer together until about 45 million years ago, in early

Tertiary time; furthermore, Australia was then probably about 15° further south than it is today. At this time, both Australia and Antarctica appear to have had a temperate climate, as shown by their fossil floras: in Antarctica, for example, the fossil remains include southern beech (*Nothofagus*) and conifers. Extensive glaciation of Antarctica apparently did not begin much before about five million years ago (Denton *et al.* 1971). Thus, before the southern land masses started to drift apart, the evolution of the flowering plants was already well established and there would have been the opportunity for interchanges of plant and animal species between different areas. Shortly we shall examine the flora in more detail for evidence of such exchanges.

The contiguities of the southern land masses in the early Tertiary have left their broad impress on the character of what is now the Kosciusko alpine flora. However, it is the more recent events of the Pleistocene and its associated cold periods which have been most important in the

Granite cliffs and crags of the Blue Lake cirque.

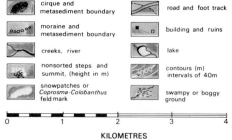

cirque and
metasediment boundary

road and foot track

moraine and
metasediment boundary

building and ruins

creeks, river

lake

nonsorted steps and
summit, (height in m)

contours (m)
intervals of 40m

snowpatches or
Coprosma-Colobanthus
feld mark

swampy or boggy
ground

Fig. 2 Major glacial and periglacial features
of part of the Kosciusko alpine area.

0   1   2   3   4

KILOMETRES

development of the present landscape with its many modifications which constitute the habitats for present-day species. Let us look more closely, then, at what we know of the last part of the Pleistocene period, as it affected the Kosciusko area (Fig. 2).

From studies in the northern hemisphere, the existence of several glaciations has been established, during which temperatures were lower than they are today. But whether there were also several glaciations at Kosciusko, more or less synchronous with those in the northern hemisphere, remains to be determined. All we can say for certain at present is that the last major glaciation in the north was also represented at Kosciusko. The technique of carbon-14 dating of organic materials, including peats and wood remains, associated with glacial and periglacial features, has revealed the following main sequence of events (Costin 1972). Many details remain to be filled in, but the broad outline now seems clear. Key dates on which the following interpretation depends are illustrated in Fig. 3.

Fig. 3 Diagrammatic view of part of the Kosciusko landscape showing some sites from which the glacial and post-glacial history of the area has been established. Details of the sites (numbered 1–13), including carbon-14 dates (in years Before Present) of important features, are shown in the corresponding panels.

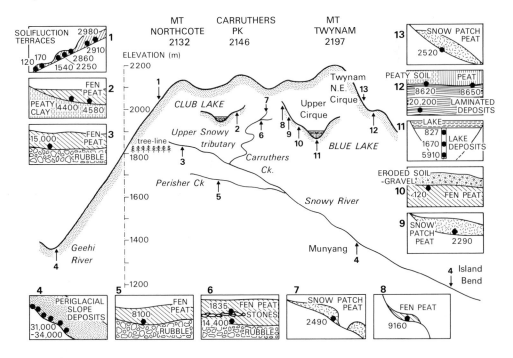

The steep slopes of the ridge between Mueller's Peak and Mt Townsend have been subjected to extensive glaciation and weathering as shown here by the ice-smoothed and ice-plucked boulders and platforms and the rocky screes. Tall heath communities are common in these sites while bog vegetation often surrounds small moraine-dammed ponds.

Alternate freeezing and thawing of soil and rock materials result in their downslope movement by processes termed solifluction. There are many solifluction features at Kosciusko, including the terraces shown here on Mt Northcote. Some of these terraces are between 1500 and 2500 years old, on the evidence of the carbon-14 ages of the fossil shrubs and other vegetation material that are buried beneath them; a lowering in mean annual temperature of about 2–3°C compared with present conditions could have caused terrace formation. Although well-preserved terraces are now restricted to a few peaks at Kosciusko, they apparently developed on a much wider scale in the past as can be appreciated by the extensive 'rippled' appearance of much of the alpine country. This rippling is most apparent under certain conditions such as in the late afternoon light, as shown here on the slopes of Mt Northcote with the so-called Railway Moraine in the foreground.

Our evidence commences with the well-developed periglacial screes or blockstreams which must have required much lower temperatures to produce the extensive shattering and downslope movement of rocks which have occurred. On the Toolong Range to the north of Kosciusko, a blockstream has overrun a stump of a temperate-climate southern beech resembling *Nothofagus cunninghamii*, still common in Tasmania but restricted in Victoria and now absent from New South Wales (Caine and Jennings 1968). The stump has been dated at about 35 000 years old and therefore gives a maximum age for the onset of the colder conditions. The maximum age of related periglacial slope deposits – as dated at Geehi, Munyang and Island Bend – is broadly similar at about 32 000 years. The development of these slope deposits, like the blockstreams, requires frozen subsoils over which topsoil material, saturated from melting snow and ice, can move. The development of these conditions on an extensive scale in the Kosciusko area implies a lowering in mean annual temperature of about 9–10°C compared with present temperatures. The depth of the slope deposits indicates that the cold climate continued for a long time.

Solifluction terraces in tall alpine herbfield, Carruthers' Peak, now covered by the *Poa* grass component of *Celmisia–Poa* tall alpine herbfield.

The pattern of solifluction terraces on Mt Northcote is visible on the large snow patch.

Wherever conditions favoured the accumulation and persistence of snow and ice, glaciers also developed, as on parts of the Main Range. For example, there are lake deposits resembling varves in the north-eastern cirque of Mt Twynam which are approximately 20 000 years old, indicating that at least cirque glaciers or large semi-permanent snow patches were in existence at that time.

The first evidence of general improvement of the climate is in the form of old peats overlying glacial or periglacial rubble in the upper Snowy Valley and the Carruthers' Creek area. These peats are about 15 000 years old and show that these and similar parts of the Kosciusko area were by then becoming sufficiently ice-free to permit plant growth. However, in leeward sites favourable for the accumulation and persistence of snow, small cirque glaciers may well have persisted for several thousand years longer since peats in these sites are not more than 9000–10 000 years old.

Temperatures then increased to at least present-day levels to promote extensive plant growth with peat formation and soil development, except for a brief period between about 3000 and 1500 years ago when there was a return to slightly colder conditions. During this so-called 'little ice age' mean annual temperatures were apparently at least 3°C lower than at present, sufficient to cause local subsoil freezing and slope instability, and an increase in snow-patch activity, at least at the highest levels. Many of the soil terraces which are such a conspicuous feature of the areas of altered sedimentary rocks between Mt Northcote and Mt

Glacial action at Kosciusko in the past was also accompanied and followed by extensive periglacial activity due to the effect of deep freezing and seasonal thawing of exposed soil and rock surfaces not insulated by a permanent cover of ice and snow. Thus, many exposed peaks underwent severe shattering, and boulders and other frost debris accumulated around them. This debris was also affected by deep freezing and seasonal thawing which, on sloping ground, resulted in its slow downslope movement. The periglacial blockstream shown here between Mt Stilwell and the Snowy River was formed by these processes. Water from subsequent snow-melt and rains has washed out much of the finer soil material which originally formed part of the blockstream, leaving only the larger boulders.

Twynam were formed at this time. From the fact that a fall in mean annual temperature of only 3°C would be sufficient to induce such landscape instability, it can be appreciated that parts of the high country are now on a knife edge between stability and erosion.

In summary, then, we now see in the Kosciusko alpine area the remains of an ancient peneplain surface, with its two main rock types of granite and altered sediments. This surface has been uplifted and variously dissected by different forms of erosion. On the steep western slopes vigorous river action has cut back deep valleys almost to the Great Dividing Range itself, whereas on the more gently sloping eastern side river erosion has been slower and the headwaters of some of the streams (notably the upper Snowy River) are still in their ancient pre-uplift condition. Glacial and periglacial erosion and deposition have diversified the landscape further. This landscape now interacts with the climate to produce still greater variation in the alpine environment. In particular, the movement of the moisture-bearing subantarctic weather systems more or less at right angles across the long north–south axis of the Main Range produces large differences in the amount and persistence of precipitation on different aspects – from relatively light and non-persistent snow cover on the wind- and sun-exposed west-facing slopes to deep semi-permanent snow patches in sheltered leeward sites. Therefore, we now find a very diverse group of environments, or habitats, for plant growth in the Kosciusko alpine area. A correspondingly wide range of distinctive plant communities (or vegetation types) has developed in response to this diversity (see vegetation maps, pp. 13–15).

HUMAN HISTORY AND THE KOSCIUSKO FLORA   Man's association with the Kosciusko alpine flora begins with the Aborigines. Carbon-14 dates of charcoals associated with sites of Aboriginal occupation show that in many inland and coastal parts of Australia Aborigines had been present for at least 20 000–30 000 years before the arrival of white settlers (Mulvaney and Golson 1971). However, no sites approaching this age have yet been discovered from the Australian high country; in fact, even in the adjacent tableland areas, there is no evidence of Aboriginal occupation beyond the last 4000 years (Flood 1973). This is perhaps not surprising in view of the evidence just summarized that cold, inhospitable glacial and periglacial conditions existed in the Kosciusko area from about 32 000 to at least 15 000 years ago, and that substantial warming up may not have occurred until about 9000 years ago.

Furthermore, the records of explorers and early settlers indicate that the alpine areas and their environs were not permanently occupied by the Aborigines. In contrast to the more extensive alpine and arctic-alpine areas of the northern hemisphere, which have abundant bird and mammal faunas, the Australian high country is relatively poor in game. Hence the Aborigines had no reason to live in the vicinity of Kosciusko or to burn off its vegetation – as they did at lower elevations – to attract game to the more palatable areas of regrowth. Nor was there an abundant source of food plants among the alpine flora. The main attraction of the Kosciusko alpine area was the annual migration of the Bogong Moth (*Agrotis infusa*). These nutritious and evidently palatable moths swarm in vast numbers to the mountains, where they rest during the summer among rocks and in crevices before flying back to lower country in autumn to breed (Common 1954). Each summer the Aboriginal tribes moved up from their lower camps to the Kosciusko area and adjacent mountains to feast on these moths, which they collected and cooked by lighting small fires among the rocks (Flood 1973). But apart from this activity, primitive man and wild herbivores appear to have had no effect on the evolution of the Kosciusko alpine flora. Therefore the flora was in no sense 'pre-adapted' to the pastoral practices of European man.

Precisely when the Kosciusko area was first visited by white men is not certain. The Polish-born explorer, Strzelecki, is usually credited with the first ascent of Mt Kosciusko. In 1840 he approached the area from the Murray side and made his ascent along what is now known as Hannels Spur. It is also possible that another Polish-born explorer, Lhotsky, reached the summit as early as 1834 (Jeans and Gilfillan 1969) but the most recent research (Andrews 1973) on Lhotsky's diaries, maps and most likely route suggests rather that Lhotsky reached an alpine peak (known as Mt William IV or Mt Terrible) well to the south of the Kosciusko Plateau, at the headwaters of the Mowamba, Jacobs and Thredbo Rivers. Perhaps both explorers were preceded by local stockmen since, by the 1830s, all of the main pastoral runs on either side of the mountains had been occupied (Hancock 1972) and the squatters were always on the lookout for new pastures. James Spencer, the well-known squatter at Waste Point near Jindabyne, apparently was the first to use the Kosciusko Plateau (known as his Excelsior Run) for grazing livestock. He came to know the area so well that he acted as guide to many expeditions.

Botanists and naturalists were also eager to explore and to collect specimens of the flora. Outstanding among these was Ferdinand Mueller (later

About the turn of the century, and especially during drought years, large numbers of sheep and cattle were sent to the mountains for summer pasture, initiating some of the erosion damage which is still a problem today. This old photograph shows a mob of cattle grazing the locally restricted but palatable fen vegetation of *Carex gaudichaudiana* around Club Lake.

Baron von Mueller). Mueller was undoubtedly the greatest of all Australian botanists and ranks among the great botanists of all time. The following notes are based on a biography by Margaret Willis (1949).

Mueller was born in Rostock, Germany, in 1825. Apprenticed to a pharmacist, he became a qualified chemist in 1846, but, from childhood, his passion was collecting and identifying plants and he soon became known to many notable botanists in Europe. Following the death of his parents, Mueller was left with the responsibility of his two sisters, one of whom suffered from consumption. The decision was made to leave Europe and travel to Adelaide, South Australia, both because of the healthier climate there and because of the opportunities it would provide for Mueller to collect a still largely unknown flora.

The Muellers arrived in Adelaide in 1847. Mueller obtained a position as an assistant chemist, but lost no opportunity to study the flora. In a still little-known country such studies often involved exploration as much as collecting plants. By this time Mueller had become known to the eminent British botanist at Kew, Sir William Hooker, who recommended Mueller for the position of Government Botanist when it was created in Melbourne in 1852. Mueller eagerly accepted the position. The Melbourne Botanic Gardens as most of us know them today reflect the planning, landscaping and imaginative planting that Mueller commenced and William Guilfoyle completed.

But Mueller's main contributions were in scientific botany. Immediately after his appointment in 1853, he threw himself into the task of botanical collecting. This involved absences of weeks and months at a time usually with no other company than a pack-horse for his supplies and botanical collections and a saddle-horse for himself. Mueller's first expedition in Victoria was to the Buffalo Plateau and Mt Buller where he made his initial acquaintance with Australian alpine plants. His second expedition, in 1853–54, also included visits to the high country near the Victoria–New South Wales border. On the third expedition, in 1854–55, he successfully explored and collected on Mt Bogong and Mt Feathertop, two of Victoria's highest peaks. By New Year's day 1855 he was collecting on the Main Range around Kosciusko; Mueller's Peak, between Mt Kosciusko and Mt Townsend, commemorates this achievement. By now Mueller felt justified in writing to Sir William Hooker (Willis 1949, p.28), 'After having traversed now the main chains of the Snowy Mountains in so many directions I am led to believe that the plants mentioned in this and the two previous letters, and those mentioned in my reports, comprehend almost completely the Alps flora of this continent...'. Margaret Willis has pointed out, 'So complete and thorough had his collecting been on these three journeys that there was little chance for the hopeful

explorers who followed after him to add to his discoveries'. Although
experience has shown this to be somewhat of an over-statement, a glance
at the names of the authors of the species listed in the botanical descriptions
gives some idea of the major importance of Mueller's contribution to
the knowledge of the alpine flora.

But Mueller's contributions went further than his own botanical exploration and collecting. He regularly dispatched specimens and scientific
notes to Kew Herbarium and maintained a prolific correspondence with
Sir William Hooker, his botanist son Joseph, and George Bentham,
another famous Kew botanist. The collections with Mueller's notes and
correspondence not only provided much of the basis of the seven-volume
Flora Australiensis completed by Bentham between 1863 and 1878, but
helped unravel the geographical affinities of the Australian flora in relation
to the floras of other parts of the world. These plant–geographical relationships were brilliantly analysed by Joseph Hooker in his monumental
Introductory Essay to Flora Tasmaniae (1860), to which we have already
referred. Some of the alpine species were found to have Australian affinities, having been derived from ancestral lowland types, as might have
been expected. But for others the relationships were with groups of plants
in Tasmania, New Zealand, other southern land masses and even parts
of the northern hemisphere.

Although Bentham and Hooker fully acknowledged Mueller's contributions to the *Flora Australiensis*, Mueller's greatest disappointment was
that he was unable to write the *Flora* himself. The title of hereditary
baron bestowed on him by the King of Württemberg and scientific
honours throughout the world were relatively unimportant to Mueller
compared with the *Flora*. Mueller died in 1896, a disappointed and somewhat embittered man following years of increasing administrative worries
in the running of the Melbourne Botanic Gardens. *By Their Fruits*, the
title of Mueller's biography, summarizes better than any other phrase
the tremendous contribution he made to botanical knowledge, including
knowledge of the alpine flora.

In Mueller's day the Australian flora excited the interest of scientists
throughout the world. But at the time in Australia, with its pastoral econ-
omy, the main interest in native plants was as feed for livestock. By the
end of the century summer grazing of sheep and cattle with burning
of the vegetation to promote fresh regrowth had become a well-
established practice at Kosciusko, having been formalized by the N.S.W.
Department of Lands in 1889 with the introduction of the snow lease
system of land tenure. As early as 1893 Helms warned against the effects
of burning off in reducing the plant cover and thereby promoting soil
erosion. In 1898 Maiden, then Governnment Botanist in New South

Wales and interested in the Kosciusko flora, commented on the adverse effects of burning on the scenery and vegetation. But these warnings were lost in the clamour for high-country grazing areas during the disastrous drought of 1890–1901. In these earlier years stockmen remained with their sheep and cattle during much of the snow-free season, depending on pack-horses and bullock teams for the transport of their own rations and of rock-salt for their livestock.

Other uses of the Kosciusko area also developed. Tourist and recreational interest rapidly increased after the completion of the Kosciusko Hotel and the 53-km road from Jindabyne to Mt Kosciusko in 1909. Concern for the area as a water catchment can be said to have commenced in the 1920s during the construction of the Hume Dam, collecting waters from the western slopes of Kosciusko for subsequent use in irrigation areas in New South Wales, Victoria and South Australia. This use was to culminate with the Snowy Mountains Scheme, commencing in 1948 and not completed until 1972. Appreciation of the wilderness and scientific values of the Kosciusko alpine area was also growing rapidly. This growth of different interests in relation to the one limited area was bound to produce divergent opinions on how the area should and should not be used.

In 1932 a report by Byles to the Commonwealth Forestry Bureau described widespread damage to the vegetation and soils in parts of the Upper Murray catchment on the western slopes of Kosciusko. The Soil Conservation Service of New South Wales, set up in 1938, formed the same conclusion and realized that catchment protection measures were urgently needed. Private individuals and societies – notably the National Parks and Primitive Area Council of New South Wales – also pressed for the protection and better management of the Kosciusko environment. These and similar pressures led to the passing of the Kosciusko State Park Act of 1944, under which an area of about one and a quarter million acres (approx. 5000 sq km) of mountainous country centred on Kosciusko was proclaimed a Park under the control of a specially appointed Trust. At the same time, the Kosciusko alpine area itself was withdrawn from grazing in the interests of minimizing further damage to this part of the catchment area. The Act also provided for the setting aside of a Primitive Area, although none was designated at the time. Concern for the condition of the mountain catchments increased with the development of the Snowy Mountains Scheme, leading to further withdrawals of snow leases and improved fire protection. By 1958, all areas in the Park above approximately 4500 ft (1370 m) had been withdrawn from grazing.

Meanwhile the Snowy Mountains Scheme was progressing rapidly and was itself threatening scientific and wilderness values in some areas of

During the early period of snow-lease grazing, stockmen remained in the mountains to shepherd their sheep and cattle for most of the grazing season and supplies were brought up by pack horse and bullock dray. This early photograph shows a bullock team approaching Kosciusko along the original track on the Crackenback Range overlooking the Thredbo River, before the existing summit road was constructed in 1909. Already extensive replacement of the snow grass sward by bare ground and inter-tussock species had occurred.

the Park, especially in the Kosciusko alpine area itself. Eventually yielding to strong scientific and public opinion, the Kosciusko State Park Trust in 1963 declared as a Primitive Area the Kosciusko alpine area and its steep western slopes, thus ensuring protection of the flora and fauna and most of their associated environments from any major man-made disturbances. With the establishment of the National Parks and Wildlife Service in 1967, the Kosciusko area came under the control of the Service, and the Kosciusko Primitive Area has now received new status as one of the main Outstanding Natural Areas recognized in the Plan of Management of the Park (National Parks and Wildlife Service of N.S.W. 1974). Thus a large measure of protection of the Kosciusko alpine flora has now been achieved. As we shall see later, many alpine species which became rare during the period of grazing and burning are now making a spectacular recovery, with a massed flowering in summer which is surpassed in few other parts of the world.

Although grazing, burning and hydroelectric works no longer threaten the Kosciusko alpine area, new problems are arising. The widening of roads and the construction of parking areas to cope with increased motor traffic involve destruction and disturbance of the vegetation, both directly and from the accelerated run-off of rain water and melted snow, as does the use of four-wheel drive and over-snow vehicles on the Main Range itself for some park management and soil conservation purposes. Foot traffic along walking tracks is also causing local damage to some of the most restricted and fragile plant communities (Wimbush and Costin 1973). We believe these new human pressures not only require further control, but in some cases re-location and restriction of use. We hope the public itself will support such management policies as it better understands and appreciates the Kosciusko environment and its unique alpine flora.

The Plants and Plant Communities

The Kosciusko alpines number approximately 200 species and well-defined subspecies and varieties, as listed in Table 1. It will be noted that some of the species have yet to be described, and future workers are sure to split up some of the broader species into smaller taxa. Some of these taxa will possibly prove to be new and endemic species (see below). At least 27 alien or introduced plants have also become naturalized in the flora (Table 2) compared with only 6 recorded in 1951 (Costin 1954). The number of naturalized aliens may well be increasing further, owing to the extensive use of introduced grasses and clovers and pasture-hay mulches in soil conservation work along the Main Range.

Of the native plants listed about 30 are exclusively alpine at Kosciusko and at least 21 are endemic, that is, they are known to occur only in the Kosciusko area (Table 3). Sixty-one plants are the only representatives of their genus and 20 the only representatives of their family to be found above the treeline. The best-represented families are Compositae (daisies), Gramineae (grasses) and Cyperaceae (sedges), followed by Umbelliferae (carrot family), Ranunculaceae (buttercups, etc.), Juncaceae (rushes) and Epacridaceae (heaths).

In addition to the taxonomic classification of plants into species, genera, families and higher groupings, there are many other criteria used in considering plants. These include the related morphological attributes of growth form, life form and habit, physiological characteristics and geographic affinities.

We have noted already that alpine species, by definition, do not include trees. The main growth forms are perennial herbs and shrubs typically not more than 1 m tall. Growth forms can be further subdivided into so-called life forms on the basis of the position of the renewal buds on the plant concerned. Devised by the Scandinavian botanist Raunkiaer (1934), the life form characteristics of a flora are considered to reflect the adaptation of the species to withstand the unfavourable season. Plants with renewal buds close to the ground, either just above or below it, are thought to be the types best adapted to cold. They include shrubs and dwarf shrubs, perennial herbs and swamp plants. The renewal buds

TABLE 1   Native species of the Kosciusko alpine flora, their typical growth forms, and main habitats

Habitats: B, bog; F, fen; H, heath; FM, feldmark;
STG, sod tussock grassland; SAH, short alpine herbfield;
TAH, tall alpine herbfield.
‡ Species endemic to Kosciusko

Species	Common name	Growth form	Main habitats
PTERIDOPHYTA			
Lycopodiaceae			
Huperzia selago	Fir Clubmoss	Fern-like	B
Lycopodium fastigiatum	Mountain Clubmoss	Fern-like	B, TAH, H
Grammitidaceae			
Grammitis armstrongii		Fern	TAH (*Brachycome–Danthonia*)
Aspleniaceae			
Asplenium flabellifolium	Necklace Fern	Fern	TAH
Blechnaceae			
Blechnum penna-marina	Alpine Water-fern	Fern	TAH
Aspidiaceae			
Polystichum proliferum	Mother Shield-fern	Fern	TAH
Athyriaceae			
Cystopteris fragilis	Brittle Bladder-fern	Fern	TAH (*Brachycome–Danthonia*)
GYMNOSPERMAE			
Podocarpaceae			
Podocarpus lawrencei	Mountain Plum Pine	Shrub	H
ANGIOSPERMAE: Monocotyledoneae			
Gramineae			
Agropyron velutinum	Velvet Wheat-grass	Grass	TAH, STG
Agrostis meionectes		Grass	STG, SAH, F, B
A. muellerana		Grass	TAH, STG, F, B, FM (*Epacris–Chionohebe*)
A. parviflora		Grass	STG, F, B
A. venusta		Grass	TAH, STG
‡*Chionochloa frigida*	Ribbony Grass	Grass	TAH
Danthonia alpicola	Crag Wallaby-grass	Grass	TAH (*Brachycome–Danthonia*)
D. nivicola	Snow Wallaby-grass	Grass	STG, TAH, F
D. nudiflora	Alpine Wallaby-grass	Grass	STG, TAH, F
Deschampsia caespitosa	Tufted Hair-grass	Grass	STG
Deyeuxia carinata		Grass	STG, F, B, H
D. crassiuscula		Grass	TAH, STG
D. monticola		Grass	STG, TAH
D. affinis		Grass	SAH
Erythranthera australis		Grass	SAH
E. pumila		Grass	FM (*Epacris–Chionohebe*)
Hierochloe submutica	Alpine Holy Grass	Grass	TAH
Poa costiniana	Prickly Snow Grass	Grass	STG, F, B, TAH, H
P. fawcettiae	Smooth-blue Snow Grass	Grass	TAH, STG
P. hiemata	Soft Snow Grass	Grass	TAH
P. saxicola	Rock Poa	Grass	TAH
Trisetum spicatum	Bristle-grass	Grass	TAH, STG
Cyperaceae			
Carex breviculmis		Sedge	TAH, STG

Species	Common name	Growth form	Main habitats
Cyperaceae (*continued*)			
C. cephalotes		Sedge	F, B, SAH, STG
C. curta		Sedge	F, B, TAH, STG
C. echinata	Star Sedge	Sedge	SAH, TAH
C. gaudichaudiana		Sedge	F, B, STG
C. hebes		Sedge	TAH, STG
C. hypandra		Sedge	F, B
C. jackiana		Sedge	SAH, B, F, STG
Carpha alpina	Small Flower-rush	Sedge	STG, F, B
C. nivicola	Broad-leaf Flower-rush	Sedge	F, B
Oreobolus distichus	Fan Tuft-rush	Sedge	B, SAH, TAH
O. pumilio	Alpine Tuft-rush	Sedge	SAH, F
Schoenus calyptratus	Alpine Bog-rush	Sedge	SAH
Scirpus aucklandicus		Sedge	SAH, B, F
S. crassiusculus		Sedge	SAH, TAH
S. habrus		Sedge	B
S. montivagus		Sedge	SAH, F, STG
S. subtilissimus		Sedge	SAH
Uncinia compacta		Sedge	TAH, STG
U. flaccida	Mountain Hook-sedge	Sedge	TAH, STG
?U. sinclairii		Sedge	STG
U. sp.		Sedge	TAH, STG
Restionaceae			
Empodisma minus	Spreading Rope-rush	Rush	B, STG, TAH
Juncaceae			
Juncus antarcticus	Cushion Rush	Rush	SAH, B
J. falcatus	Sickle-leaf Rush	Rush	F, B
J. sp.		Rush	B, F, SAH
‡Luzula acutifolia subsp. nana		Rush	SAH
L. alpestris		Rush	STG, TAH
L. atrata		Rush	B, F
L. australasica		Rush	B
L. novae-cambriae		Rush	TAH, H
L. oldfieldii subsp. dura		Rush	FM
Liliaceae			
Astelia alpina	Pineapple-grass	Herb	B, TAH
‡A. psychrocharis	Kosciusko Pineapple-grass	Herb	B, TAH
Dianella tasmanica	Tasman Flax-lily	Herb	TAH, H
Herpolirion novae-zelandiae	Sky Lily	Herb	STG
Orchidaceae			
Caladenia lyallii	Mountain Caladenia	Herb	B
Prasophyllum alpinum	Alpine Leek-orchid	Herb	B, TAH, STG
P. suttonii	Mauve Leek-orchid	Herb	B, TAH, STG

ANGIOSPERMAE: Dicotyledoneae

Proteaceae			
Grevillea australis	Alpine Grevillea	Shrub	H

Species	Common name	Growth form	Main habitats
Proteaceae (*continued*)			
G. victoriae	Royal Grevillea	Shrub	H
Orites lancifolia	Alpine Orites	Shrub	H
Santalaceae			
Exocarpos nanus	Alpine Ballart	Dwarf shrub	H, TAH
Portulacaceae			
Neopaxia australasica	White Purslane	Herb	SAH, TAH
Caryophyllaceae			
Colobanthus affinis	Alpine Colobanth	Herb	TAH, FM
‡*C. nivicola*	Soft Cushion-plant	Cushion	FM (*Coprosma–Colobanthus*), SAH, TAH
‡*C. pulvinatus*	Hard Cushion-plant	Cushion	FM (*Epacris–Chionohebe*), TAH
Scleranthus biflorus	Two-flowered Knawel	Mat/Cushion	TAH, STG
S. singuliflorus	One-flowered Knawel	Herb/ Cushion	TAH, STG, FM (*Epacris–Chionohebe*)
Stellaria multiflora	Rayless Starwort	Herb	TAH
Ranunculaceae			
Caltha introloba	Alpine Marsh-marigold	Herb	SAH
‡*Ranunculus anemoneus*	Anemone Buttercup	Herb	FM (*Coprosma– Colobanthus*), TAH, SAH
‡*R. dissectifolius*		Herb	TAH, STG, H
R. graniticola	Granite Buttercup	Herb	TAH, STG
R. gunnianus	Gunn's Alpine Buttercup	Herb	TAH, STG
R. millanii	Dwarf Buttercup	Herb	STG, F, B
R. muelleri var. *muelleri*	Felted Buttercup	Herb	TAH, STG
‡*R. muelleri* var. *brevicaulis*		Herb	FM (*Epacris–Chionohebe*)
‡*R. niphophilus*	Snow Buttercup	Herb	SAH, TAH
Winteraceae			
Tasmannia xerophila	Alpine Pepper	Shrub	H
Cruciferae			
Cardamine sp.		Herb	STG, TAH
C. sp.		Herb	SAH, TAH (*Brachycome–Danthonia*)
Droseraceae			
Drosera arcturi	Alpine Sundew	Herb	B, SAH
Crassulaceae			
Crassula sieberana		Herb	TAH
Rosaceae			
Acaena sp.		Subshrub	TAH, H, STG
? *Alchemilla xanthochlora*	Lady's Mantle	Herb	TAH
Papilionaceae			
Hovea purpurea var. *montana*	Alpine Hovea	Shrub	H
Oxylobium alpestre	Mountain Shaggy-pea	Shrub/ Subshrub	H
O. ellipticum	Golden Shaggy-pea	Shrub/ Subshrub	H

Species	Common name	Growth form	Main habitats
Geraniaceae			
Geranium antrorsum	Rosetted Crane's-bill	Herb	TAH, STG, B
G. potentilloides var. *abditum*		Herb	TAH, STG, H
G. potentilloides var. *potentilloides*		Herb	TAH, STG
Rutaceae			
‡*Phebalium ovatifolium*	Ovate Phebalium	Shrub	H
Stackhousiaceae			
Stackhousia pulvinaris	Alpine Stackhousia	Mat	STG, TAH
Violaceae			
Hymenanthera dentata var. *angustifolia*		Subshrub	H
Viola betonicifolia subsp. *betonicifolia*	Showy Violet	Herb	TAH, STG
Thymelaeaceae			
Drapetes tasmanicus		Dwarf shrub	FM (*Epacris–Chionohebe*)
Pimelea alpina	Alpine Rice-flower	Dwarf shrub	TAH, STG, H
P. axiflora var. *alpina*		Subshrub	H, TAH, STG
P. ligustrina		Shrub	H
Myrtaceae			
Baeckea gunniana	Alpine Baeckea	Shrub	B, H
B. utilis	Mountain Baeckea	Shrub/ Subshrub	H, B
Kunzea muelleri	Yellow Kunzea	Shrub/ Subshrub	H, STG
Onagraceae			
Epilobium gunnianum	Gunn's Willow-herb	Herb	TAH, STG, B, H
E. sarmentaceum	Mountain Willow-herb	Herb	TAH, STG
E. tasmanicum	Snow Willow-herb	Herb	SAH, FM (*Coprosma– Colobanthus*)
Haloragaceae			
Gonocarpus micranthus subsp. *micranthus*	Creeping Raspwort	Herb	F, B, STG
G. montanus		Herb	TAH
Myriophyllum pedunculatum	Mat Water-milfoil	Herb	F
Umbelliferae			
Aciphylla glacialis	Mountain Celery	Herb	TAH
A. simplicifolia	Mountain Aciphyll	Herb	TAH, STG
‡*Dichosciadium ranunculaceum* var. *ranunculaceum*		Herb	SAH, B, TAH
Diplaspis hydrocotyle	Stiff Diplaspis	Herb	SAH, B, TAH
Oreomyrrhis brevipes	Rock Carraway	Herb	TAH, FM (*Epacris–Chionohebe*)
O. ciliata	Bog Carraway	Herb	B
O. eriopoda	Australian Carraway	Herb	TAH, H
O. pulvinifica	Cushion Carraway	Herb	SAH, TAH
Oschatzia cuneifolia	Wedge Oschatzia	Herb	TAH, B
Schizeilema fragoseum	Alpine Pennywort	Herb	TAH
‡*Gingidia algens*		Herb	TAH

Species	Common name	Growth form	Main habitats
Epacridaceae			
Epacris glacialis		Shrub/ Subshrub	H, B, STG, TAH
E. microphylla	Coral Heath	Shrub/ Subshrub	FM (*Epacris–Chionohebe*), H, B
E. paludosa	Swamp Heath	Shrub	B, H
E. petrophila	Snow Heath	Shrub/ Subshrub	FM (*Epacris–Chionohebe*)
Leucopogon montanus	Snow Beard-heath	Shrub	H
Pentachondra pumila	Carpet Heath	Mat/Subshrub	H, TAH, STG, FM (*Epacris–Chionohebe*)
Richea continentis	Candle Heath	Subshrub	B
Gentianaceae			
Gentianella diemensis	Mountain Gentian	Herb	STG, TAH
Boraginaceae			
Myosotis australis (Kosciusko form)		Herb	TAH
Labiatae			
Prostanthera cuneata	Alpine Mint-bush	Shrub	H
Scrophulariaceae			
Chionohebe densifolia		Dwarf shrub	FM (*Epacris–Chionohebe*)
‡*Euphrasia alsa*	Dwarf Eye-bright	Herb	FM (*Epacris–Chionohebe*), STG, TAH
E. collina subsp. *diversicolor*		Herb	TAH, STG, H
‡*E. collina* subsp. *glacialis*		Herb	STG, SAH, F, TAH
‡*E. collina* subsp. *lapidosa*		Herb	FM (*Epacris–Chionohebe*)
Veronica serpyllifolia	Thyme Speedwell	Herb	STG, B
Plantaginaceae			
Plantago alpestris		Herb	STG, TAH
P. euryphylla		Herb	STG, TAH
P. glacialis	Small Star Plantain	Herb	SAH, B
P. muelleri	Star Plantain	Herb	SAH, B
Rubiaceae			
Asperula gunnii	Mountain Woodruff	Herb	TAH, STG, H
A. pusilla	Alpine Woodruff	Herb	TAH, STG, H
‡*Coprosma* sp.		Mat/ Subshrub	FM (*Coprosma–Colobanthus*)
Nertera depressa		Herb	STG, F, B
Campanulaceae			
Wahlenbergia ceracea	Waxy Bluebell	Herb	TAH, STG
W. gloriosa	Royal Bluebell	Herb	TAH
Lobeliaceae			
Pratia surrepens	Mud Pratia	Herb	F, B, STG
Goodeniaceae			
Goodenia hederacea var. *alpestris*		Herb	TAH
Stylidiaceae			
Stylidium graminifolium	Grass Trigger-plant	Herb	TAH, STG, H, B

Species	Common name	Growth form	Main habitats
Compositae			
Abrotanella nivigena	Snow-wort	Mat/Herb	SAH
Brachycome sp.		Herb	TAH, STG, FM (*Epacris–Chionohebe*)
B. nivalis var. *alpina*		Herb	SAH, TAH
B. nivalis var. *nivalis*	Snow Daisy	Herb	TAH (*Brachycome–Danthonia*)
B. obovata		Herb	TAH
B. scapigera	Tufted Daisy	Herb	TAH, STG, H, B
‡*B. stolonifera*		Herb	SAH, TAH, STG
B. tenuiscapa var. *tenuiscapa*		Herb	STG, TAH
Celmisia sp.	Silver Snow Daisy		TAH, STG, B
Cotula alpina	Alpine Cotula	Herb	B, TAH, STG
‡*Craspedia leucantha*		Herb	SAH, TAH
‡*C.* sp. A		Herb	STG
C. sp. B		Herb	TAH (*Brachycome–Danthonia*)
‡*C.* sp. C		Herb	TAH, STG
C. sp. D		Herb	TAH, STG
C. sp. E		Herb	TAH, STG
C. sp. F		Herb	TAH
Erigeron pappocromus (form A)		Herb	STG, TAH
E. pappocromus (form B)		Herb	STG, TAH
E. pappocromus (form C)		Herb	B
‡*E. setosus*		Herb	SAH
Ewartia nubigena	Silver Ewartia	Mat/Subshrub	FM (*Epacris–Chionohebe*), TAH
Gnaphalium argentifolium	Silver Cudweed	Mat/Herb	TAH, STG
G. fordianum		Herb	TAH, STG
G. nitidulum	Shining Cudweed	Mat/Herb	B, STG
G. umbricola	Cliff Cudweed	Herb	TAH
Helichrysum alpinum	Alpine Everlasting	Shrub	B, H
H. hookeri	Scaly Everlasting	Shrub	B, H
H. scorpioides	Button Everlasting	Herb	TAH
H. secundiflorum	Cascade Everlasting	Shrub	H
Helipterum albicans subsp. *alpinum*	Alpine Sunray	Mat/Herb	TAH, FM (*Epacris– Chionohebe*)
H. anthemoides	Chamomile Sunray	Herb	TAH
Lagenifera stipitata		Herb	STG, TAH
Leptorhynchos squamatus	Scaly Buttons	Herb	STG, TAH
L. squamatus (feldmark ecotype)		Herb	FM (*Epacris–Chionohebe*)
Microseris lanceolata	Native Dandelion	Herb	TAH, STG
Olearia algida	Alpine Daisy-bush	Shrub	H
O. phlogopappa var. *flavescens*		Shrub	H
O. phlogopappa var. *subrepanda*		Shrub	H
Parantennaria uniceps		Mat/Herb	SAH, STG, B
Podolepis robusta	Alpine Podolepis	Herb	TAH, STG
Senecio gunnii		Herb	TAH
S. lautus subsp. *alpinus*	Variable Groundsel	Herb	TAH, STG
S. pectinatus	Alpine Groundsel	Herb	TAH, FM (*Epacris–Chionohebe*)

TABLE 2   Naturalized species of the Kosciusko alpine flora

Species	Common name
Agrostis capillaris	Brown-top Bent
A. stolonifera	Creeping Bent
?*Alchemilla xanthochlora*	Lady's Mantle
Aphanes arvensis	Parsley Piert
Arenaria serpyllifolia	Thyme-leaved Sandwort
Bromus diandrus	Great Brome
B. hordeaceus	Barley Brome
Capsella bursa-pastoris	Shepherd's Purse
Cerastium fontanum subsp. *triviale*	Mouse-ear Chickweed
Crepis capillaris	Smooth Hawk's-beard
Dactylis glomerata	Cocksfoot
Festuca rubra	Red Fescue
Hordeum glaucum	Barley-grass
Hypochoeris radicata	Cat's-ear (Flatweed)
Juncus articulatus	Jointed Rush
Lolium perenne	Perennial Ryegrass
Myosotis discolor	Yellow and Blue Forget-me-not
Phleum pratense	Timothy Grass
Picris hieracioides	Hawkweed Picris
Poa annua	Annual Meadow-grass (Winter-grass)
P. pratensis	Kentucky Blue-grass
Rumex acetosella	Sheep Sorrel
Spergularia rubra	Sand-spurrey
Taraxacum officinale sp. agg.	Dandelion
Trifolium arvense	Hare's-foot Clover
T. pratense	Red Clover
T. repens	White Clover
Veronica arvensis	Wall Speedwell

of such species are protected from cold either by the basal leaves or surface litter or by the soil or mud. Annual plants with a short life cycle may also escape the unfavourable season by rapid growth, flowering and setting of seed.

However, there are always cases that fall between rather than within the groups being recognized. For example, the usual distinction between shrubs being woody and perennial herbs being non-woody is not always easy to sustain. Some of the species of very wind-exposed or snow-patch situations at Kosciusko (e.g. *Chionohebe* and *Drapetes*) have leaves on non-woody shoots supported by short semi-woody stems. Other intermediate forms between dwarf shrubs and herbs are the mat and cushion forms, as in *Coprosma* and some species of *Colobanthus*. These two intermediate life forms are characteristic of many alpine floras, including the alpine flora of Kosciusko (cf. Table 1).

TABLE 3   Plants endemic to the Kosciusko area†

Scientific name	Common name
Astelia psychrocharis	Kosciusko Pineapple-grass
Brachycome stolonifera	
Chionochloa frigida	Ribbony Grass
Colobanthus nivicola	Soft Cushion-plant
C. pulvinatus	Hard Cushion-plant
Coprosma sp.	
Craspedia leucantha	
C. sp. A	
C. sp. C	
Dichosciadium ranunculaceum var. *ranunculaceum*	
Erigeron setosus	
Euphrasia alsa	Dwarf Eye-bright
E. collina subsp. *glacialis*	
E. collina subsp. *lapidosa*	
Gingidia algens	
Luzula acutifolia subsp. *nana*	
Phebalium ovatifolium	Ovate Phebalium
Ranunculus anemoneus	Anemone Buttercup
R. clivicola (subalpine tract)	
R. dissectifolius	
R. muelleri var. *brevicaulis*	
R. niphophilus	Snow Buttercup

†This list is tentative only and will need to be modified when further revisionary work is carried out on critical genera.

Still further modification of the shrub life form often found in alpine areas is seen in the espalier and rock-clinging habits: *Podocarpus*, *Phebalium*, *Pentachondra*, *Grevillea* and *Kunzea* provide typical examples at Kosciusko. The rock-clinging habit is often developed on sunny and exposed rather than shady and protected rock faces, apparently reflecting the requirement of many alpine species to maximize the amount of energy available for growth during the snow-free season. The exposed rocks are effective absorbers of solar energy and thus provide locally warmer habitats for plant growth. This may also be the explanation for the upward extension of the treeline into rocky situations (Wimbush and Costin 1973).

Because of the imperfections of the various life- and growth-form classifications of plants, in Table 1 we have adopted the compromise approach of using the following general terms which will be familiar to most people: fern-like, fern, rush, sedge, grass, mat, cushion, herb, dwarf shrub, subshrub and shrub.

The floras of most other alpine areas of the world, although largely composed of species different from those at Kosciusko, show similar specialized life forms and morphological features (e.g. Billings 1973; Mark and Adams 1973; Zwinger and Willard 1972).

The seasonality of the alpine climate, with its contrasting cold winter–spring period with snow cover and the warmer summer–early autumn, is strongly reflected in the behaviour of the flora. Most of the species show a winter dormancy followed by a rapid growth period including flowering. Many species have their peak of flowering in late January and early February (e.g. *Celmisia*, *Craspedia* and *Euphrasia*); a few species including the Anemone Buttercup (*Ranunculus anemoneus*) and Alpine Marsh-marigold (*Caltha introloba*) flower earlier – soon after the snow has melted or even while still under the snow – evidently drawing

Cross section of a trunk of *Podocarpus lawrencei* (Mountain Plum Pine) from a rock-sprawling plant growing on Etheridge Range. The growth rings indicate an age of more than 170 years although the diameter of the trunk is barely 6 cm. Such long-lived species require careful preservation including protection from fire.

upon reserves of food accumulated in roots and stems during the previous season. Yet others – including the Mountain Gentian (*Gentianella diemensis*), Grass Trigger-plant (*Stylidium graminifolium*) and Waxy Bluebell (*Wahlenbergia ceracea*) – are late-flowering, not reaching their peak until late February or March. Thus, although the main flowering of the alpine flora occurs in mid summer, there is a sequence of flowering virtually throughout the whole of the snow-free season.

We have already noted that summer temperatures are relatively low and, at least for some types of plants (namely, trees and tall shrubs), are apparently inadequate for survival. The limited amount of energy available for photosynthesis and plant growth is also associated with the very slow growth rates of many alpine species; in other words, the balance between the energy accumulated by photosynthesis in summer and energy lost in respiration during the whole year is barely positive. The dwarf feldmark epacrids increase their stem diameter by as little as 0.27 mm and their length by only about 1 cm per year. For the long-lived Mountain Plum Pine (*Podocarpus lawrencei*) the rate of stem-diameter growth is similarly small in exposed sites (about 0.25 mm per annum) (Barrow *et al.* 1968; Costin *et al.* 1969).

Although plant growth generally in alpine areas appears to be limited by the shortage of energy, there may also be short periods when plants are exposed to intense insolation, at which times the ability of the leaves to keep cool may be critical for survival. Transpiration of water from the leaves is the normal cooling mechanism, but efficient transfer of heat from the leaf surface to the air is also important. In this respect species with small needle-shaped leaves, with a large surface area in relation to their volume, are more efficient in dissipating heat than larger-leaved species. The predominance of small-leaved species, notably the dwarf epacrids, in the most sun- and wind-exposed sites may in part reflect the importance of rapid energy transfer as a plant survival mechanism.

However, the physiology of alpine plants has not been adequately studied and the special mechanisms of growth and survival are, as yet, imperfectly understood. A recent review by the American ecologist Bliss (1971) summarizes the available information.

From time to time we have alluded to the apparent relationships of the Kosciusko alpine flora to floras in other parts of the world, and have proposed that these relationships indicate the former proximity or connection of land masses which are now separated. In addition to the Australian and cosmopolitan elements in the Kosciusko flora, there are elements with affinities and possible origins in the tropics, South Africa, Antarctica, New Zealand and South America. As might be expected, the proportions of species with Australian and cosmopolitan affinities are high,

4 Early summer view of the Kosciusko Main Range from the east showing the semi-permanent snow patches characteristic of the leeward slopes. Winter storms and blizzards cause heavy accumulation of snow in these sites sheltered from the prevailing westerlies. Stunted Snow Gums of the treeline are in the foreground. The vegetation is mainly tall alpine herbfield, the darker patches heath.

and the proportion of species with tropical affinities is low. Collectively, the southern elements are also strongly represented, accounting for almost 25% of the alpine flora.

We interpret the strong representation of the southern geographical elements as indicating former contiguity of an old Antarctic continent with South Africa, South America, New Zealand and Australia. But the distribution of the southern elements in the present-day mountain floras of these countries is not uniform; for example some old Antarctic types occur in one or more countries but not in others. Such apparently anomalous distributions suggest that South Africa was separated very early from the old Antarctic continent, and hence from other southern areas; and that Australia and New Zealand became separated from each other whilst they were both still connected to Antarctica and while South America and Antarctica were still close together. Long-distance transport of seeds by wind and by birds may also have occurred; some botanists (e.g. Raven 1973) argue that this could have been the main dispersal mechanism. Detailed cytological studies of the geographically related groups of species in these southern areas are required to determine the most likely sequence of plant evolution and distribution. But from the existing geological and botanical evidence we infer that former connections existed between the southern land masses and an old Antarctic continent, during which there were differential interchanges of species between various areas. Similar relationships in the fauna of the Kosciusko area, especially the insects which are the main components, are evident (Mackerras 1970).

We have now reached the point of realizing that the species of the alpine flora – and indeed the species of almost any flora – do not grow haphazardly together, but in distinctive communities of species adapted to different sets of environmental conditions or habitats, and often reflecting different origins. Our next step is to look at the plant communities of the Kosciusko alpine area and the distinctive habitats in which they occur.

THE PLANT COMMUNITIES Plants grow neither in random mixtures nor in isolation, but as members of communities. These communities have different preferences and tolerances with respect to climate, soils, aspect and competition from other species and communities. Hence any diverse area such as Kosciusko usually supports a mosaic of plant communities reflecting the main combinations of environmental conditions. The ability to recognize these communities and relate them to their environment gives a deeper understanding and appreciation of the factors determining plant distribution. We shall, therefore, outline the main plant communities for the Kosciusko alpine area and its surroundings, according

5

6

5 The rugged western slopes of the Main Range, as shown here on Sentinel Peak and Watson's Crags, support tree growth in localized sheltered sites and aspects. Above the trees are heaths and tall alpine herbfield including the tall tussock grass *Chionochloa frigida*.

6 Part of the Kosciusko plateau looking north-west from above Dead Horse Gap in the Thredbo valley, showing the sequence from the wooded subalpine tract to the snow-covered treeless alpine zone. The treeline tends to be diffuse and discontinuous in some areas, reflecting locally favourable and unfavourable conditions for tree growth. The favourable sites are generally the warmer ones such as rocky outcrops, and the unfavourable ones the cold and snowy aspects.

7 The treeline on the eastern slopes of the Kosciusko plateau is often characterized by distinct patterns of roughly parallel rows of trees at the highest elevation of tree growth.

8 Outcrops of granite provide shelter for the dwarf Snow Gums of the treeline and for *Podocarpus* heath as shown here on the Ramshead Range.

7   8

9 This old Snow Gum beyond Charlotte's Pass is near the extreme limits of tree growth. Mt Lee (left) and Carruthers' Peak (right) form part of the Main Range, from which Club Lake Creek drains into the Snowy River.

10 Vertically dipping slates and phyllites of Ordovician age form a narrow band of rocks which extends through Watson's Crags to beyond the Etheridge Range. These rocks, which are associated with some of the most spectacular scenery in the mountains, are shown here on the ridge to Sentinel Peak. An outcrop of these rocks and tall alpine herbfield with *Craspedia* in flower are in the foreground.

11 The narrow band of vertically dipping metasediments in the Kosciusko alpine area was particularly susceptible to glacial erosion, resulting in such well-developed glacial features as Club Lake cirque below Carruthers' Peak. The bright green vegetation at the head of the lake is *Carex gaudichaudiana* fen. The patches of bright green vegetation below bare rock areas are short alpine herbfield while the slope in the foreground supports tall alpine herbfield.

10

11

9

12 Granitic rocks such as those shown here on Etheridge Range are the most widespread rock type in the Kosciusko alpine area. Many periglacial block-streams originate from such outcrops.

13 Aerial view of periglacial and glacial features in the Blue Lake area. The smoothing of the upper slopes is the result of periglacial solifluction processes associated with differential freezing and thawing of the soil, in contrast with the more spectacular glacial cirques of Blue Lake and the extensive moraines derived from the excavation of the cirques. Part of this moraine (foreground) dams back the creek flowing from Blue Lake to form Hedley Tarn. Blue Lake, with an area of about 15 ha and a maximum depth of about 30 m, is the largest of the glacial lakes in the Kosciusko alpine area. Communities of heath have developed extensively on the rocky moraine.

14 The approximately vertical and horizontal jointing of the granite rocks facilitated the excavation of the deep cirque now occupied by Blue Lake, when glaciers were active during the Ice Age. The horizontal platforms in the middle distance have been formed by the pulling away of blocks of rock by ice action. The damp crevices in the cliffs support some of the rarer plants in the Kosciusko alpine area including *Gnaphalium umbricola* and *Cystopteris fragilis*.

15 Western aspects of the Main Range from the glaciated valley of Lake Albina showing Watson's Crags and Carruthers' Peak in the background. The foreground shows typically ice-plucked and smoothed rocks of this area and part of the extensive erratic-studded lateral moraine on the far bank.

14

15

16 >

16 Although most of the precipitation on the alpine area falls as snow, heavy rainstorms are common in summer and are often preceded by spectacular cloud formations as shown here near Mueller's Peak.

17 The Kosciusko alpine area contains many examples of glacial action, ranging from smooth-sided moraine-covered valleys with head-wall cirques to ice-smoothed and ice-polished pavements such as the one shown here above Lake Albina. The boulders on the pavement are glacial erratics that have been transported by ice. Numerous larger erratics are also evident on the opposite slope. The valley of Lake Albina itself is partly the result of erosion by a small valley glacier and the lake has been dammed back behind glacial moraine.

18 When the Kosciusko glaciers melted, the soil and rock material originally mixed with the ice was left behind in the form of moraine as seen here on the slopes near Lake Albina. The well-drained rocky parts of the moraine usually support heath vegetation.

17

18

19 Moraines often contained pockets of 'dead ice' which on melting caused local subsidence to form depressions or 'kettles', some now recognizable as small tarns and ponds.

20 The steep western face of the Main Range can be appreciated from the head of the Lady Northcote's Canyon where the creek leaves Lake Albina.

19

20

21 A mid-summer snowfall highlights the rippling effect of the solifluction terraces on Mt Northcote. 22 A winter view of part of the Main Range with Mt Townsend at the top left. The Blue Lake cirque is on the right with the curve of its associated terminal moraine in the foreground.

23 Winter scene in the Kosciusko alpine area showing
Carruthers' Peak, one of the highest peaks on the Main Range.
The cirque below the peak is occupied by Club Lake which,
like the creeks and the Snowy River (right foreground) at this
time of the year, is normally bridged over by ice and snow.
The projecting rocks are part of the narrow band of vertically
dipping slates and phyllites which run through the alpine area
in an approximately north–south direction.

22

23

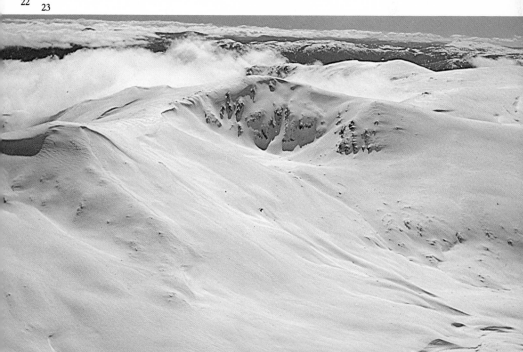

24

25

Spring and early summer scenes in the Kosciusko alpine area showing intermediate stages in the disappearance of the deep winter snow pack.
24 The Main Range with Mueller's Peak in the centre.
25 Near the source of the Snowy River.
26 Blue Lake Creek discharges snow melt from the lake.
27 Semi-permanent snow patch near Mueller's Peak with Mt Townsend behind.

26

27

28 Semi-permanent snow patch on the leeward slopes of Mt Kosciusko occupies the stony site supporting *Coprosma–Colobanthus* feldmark. The dark green patches below are short alpine herbfield, and tall alpine herbfield communities are in flower.

29 Leeward slopes of the Main Range near Carruthers' Peak showing the persistent snow patches and associated features. The apparently bare, stony areas surrounding the snow patches are occupied by sparse feldmark communities of the *Coprosma–Colobanthus* alliance and the bright green areas below, receiving abundant snow-melt water, by *Plantago–Neopaxia* short alpine herbfield. The vegetation along the ridge (left foreground) is mainly feldmark of the *Epacris–Chionohebe* alliance with tall alpine herbfield in the more sheltered sites.

28

29

30 The *Epacris–Chionohebe* feldmark alliance on a wind-exposed, west-facing slope near Carruthers' Peak. Owing to extreme exposure to the prevailing wind, the individual plants of *Epacris microphylla* tend to be regularly spaced on the stony erosion pavement. The community also is relatively snow-free and thus exposed to extremes of temperature from autumn to spring when other alpine communities are protected by snow; in summer they are exposed to drying winds. Sentinel Peak is in the background.

31 Snow patch sites which do not receive melt water from the snow patch (in contrast to the wet, short alpine herbfield areas) are occupied by communities of the *Coprosma–Colobanthus* feldmark alliance. The large green mats are *Coprosma* and the small cushions *Colobanthus*.

30

31

32

33

32 Part of Upper Snowy Valley showing the association of heath of the *Oxylobium–Podocarpus* alliance with rocky soils on the east side of the valley contrasting with the grassland and herbfield alliances on the relatively stone-free soils on the opposite side. Etheridge Range is in the background.
33 Heath of the *Oxylobium–Podocarpus* alliance is extensively developed on well-drained, steep, relatively snow-free western slopes as seen here near Mt Twynam. Such areas are the main alpine habitats for *Grevillea victoriae* and *Oxylobium alpestre*.

to the criteria of vegetation structure and floristics. The structural characteristics of vegetation are based on the growth forms of plants (e.g. trees, tall shrubs, dwarf shrubs, herbs) and their arrangement (e.g. closely spaced, in vertical layers). The floristic criteria are based on the most conspicuous and characteristic species of the community concerned. For example, herbaceous species mainly characterize herbfield and grassland vegetation, which may be further differentiated according to characteristic plant species into so-called 'associations' and groups of related associations termed 'alliances'. The main communities in the Kosciusko area as recognized by these structural and floristic criteria are listed in Table 4; their distributions are shown in the colour maps.

Reference to the colour maps and to Fig. 4 shows the relationship of the alpine plant communities as a whole to the communities at lower levels. These relationships are best appreciated on the steep, western side of Kosciusko where, over a distance of a few kilometres, there is an unparalleled sequence of lowland, mountain, subalpine and alpine environments. The lowland communities, which are relatively restricted, consist of savanna woodlands of the *Eucalyptus pauciflora –E. stellulata* (White Sally–Black Sally) alliance, occurring mainly along the lower rivers and tributary creeks; and patches of dry sclerophyll forests of the *Eucalyptus macrorhyncha–E. rossii* (Red Stringybark–Scribbly Gum) alliance on dry stony sites. Most of the mountain slopes themselves are occupied by wet sclerophyll forests. There are two main types: communities of the *Eucalyptus fastigata–E. viminalis* (Brown Barrel–Ribbon Gum) alliance between about 450 and 1000 m, grading into communities of the *Eucalyptus delegatensis–E. dalrympleana* (Alpine Ash–Mountain Gum) alliance which extends up to elevations of about 1500 m. The tallest growing forest tree is the Alpine Ash which may exceed 40 m in height. Above about 1500 m there is a fairly abrupt change from tall forests to lower-growing and more open subalpine woodland and scrub of the *Eucalyptus pauciflora* subsp. *niphophila* (Snow Gum) alliance, in which the Snow Gum is the main tree species. This change occurs at about the lower level of the winter snow line, i.e. where there is a continuous snow cover of at least one month, corresponding to a mean mid-winter temperature of less than 0°C. The Snow Gum communities extend up to about 1830 m, the exact level depending on local conditions, above which the truly alpine vegetation begins. However, as can be seen from the maps, a few scattered groups of Snow Gums extend beyond the general treeline, on locally warmer sites, as islands of subalpine vegetation within the main alpine zone.

As we have noted, the alpine vegetation itself contains no trees, being characterized by low-growing shrubs and herbs forming a variety of feldmark, heath, herbfield, grassland, fen and bog communities (Table 4).

TABLE 4   Main plant communities of Kosciusko alpine area

Formation or subformation	Alliance	Main association-dominants; and some other characteristic species	Distribution
Sod tussock grassland	[1]*Poa–Danthonia*	*Poa costiniana, Danthonia nudiflora, D. nivicola, Agrostis* spp., *Empodisma minus, Carpha alpina, Luzula alpestris, Craspedia* sp. A, *Gentianella diemensis, Euphrasia collina* subsp. *glacialis* etc.	Low-lying situations, mainly along valleys
Tall alpine herbfield	[2]*Celmisia–Poa* (including *Chionochloa frigida* association)	*Celmisia* sp., *Poa* spp., *Helipterum albicans* subsp. *alpinum, Chionochloa frigida, Aciphylla glacialis, Craspedia* spp., *Euphrasia collina* subsp. *diversicolor* etc.	Main alpine community on 'average' sites
	[3]*Brachycome–Danthonia*	*Brachycome nivalis* var. *nivalis, Danthonia alpicola, Alchemilla xanthochlora, Blechnum pennamarina, Polystichum proliferum* etc.	Restricted to rock faces, boulders and crevices
Short alpine herbfield	[4]*Plantago–Neopaxia*	*Plantago glacialis, P. muelleri, Neopaxia australasica, Caltha introloba, Ranunculus niphophilus, Oreobolus pumilio, Schoenus calyptratus, Luzula acutifolia* subsp. *nana, Oreomyrrhis pulvinifica, Dichosciadium ranunculaceum, Abrotanella nivigena, Parantennaria uniceps, Brachycome stolonifera, Erigeron setosus* etc.	Local occurrences on wet sites below snow patches, and on wet semi-bare surfaces

Alliance names according to earlier terminology (Costin 1954):		
	[1]*Poa caespitosa–Danthonia nudiflora*	[6]*Coprosma pumila–Colobanthus benthamianus*
	[2]*Celmisia longifolia–Poa caespitosa*	[7]*Carex gaudichaudiana–Sphagnum cymbifolium*
	[3]*Brachycome nivalis–Danthonia alpicola*	[8]*Epacris paludosa–Sphagnum cymbifolium*
	[4]*Plantago muelleri–Montia australasica*	[9]*Epacris serpyllifolia–Kunzea muelleri*
	[5]*Epacris petrophila–Veronica densifolia*	[10]*Oxylobium ellipticum–Podocarpus alpinus*

The feldmark communities, characterized by scattered dwarf prostrate plants some with a mat or cushion habit, occur in the most unfavourable situations for plant growth. These situations are of two main types: late-lying snowdrift areas on leeward slopes, and cold very wind-exposed ridges. The former sites are snow-covered for many months of the year, sometimes for more than a year at a time. When the snow melts it exposes rocky ground, which may then be subject to strong insolation during the day and freezing temperatures at night. At Kosciusko, the characteristic plants of these situations are the mat and cushion plants *Coprosma*

Formation or subformation	Alliance	Main association-dominants; and some other characteristic species	Distribution
Feldmark	[5]*Epacris–Chionohebe*	*Epacris microphylla, E. petrophila, Chionohebe densifolia, Euphrasia collina* subsp. *lapidosa, Ewartia nubigena, Helipterum albicans* subsp. *alpinum, Drapetes tasmanicus, Ranunculus muelleri* var. *brevicaulis, Colobanthus pulvinatus, Luzula oldfieldii* subsp. *dura, Erythranthera pumila* etc.	Very wind-exposed sites
	[6]*Coprosma–Colobanthus*	*Coprosma* sp., *Colobanthus nivicola, Ranunculus anemoneus, Epilobium tasmanicum* etc.	Local occurrences in association with snow patches
Fen	*Carex gaudichaudiana*	*Carex gaudichaudiana, C. hypandra, Danthonia nudiflora* etc.	Local occurrences in permanently wet, almost level sites
Valley bog	[7]*Carex–Sphagnum*	*Sphagnum cristatum, Carex gaudichaudiana* etc.	As above, but generally under more acid conditions
Raised bog	[8]*Epacris–Sphagnum*	*Sphagnum cristatum, Carex gaudichaudiana, Epacris paludosa, Richea continentis, Astelia* spp. etc.	Wet, acid sites, usually along edges of valleys and around hillside springs
Heath	[9]*Epacris glacialis*	*Epacris glacialis, Poa costiniana* etc.	Damp sites marginal to and usually surrounding bogs
	[10]*Oxylobium–Podocarpus* (including *Kunzea muelleri* association)	*Oxylobium ellipticum, Podocarpus lawrencei, Kunzea muelleri, Phebalium ovatifolium, Grevillea australis, Prostanthera cuneata* etc.	Common in rocky, well-drained sites

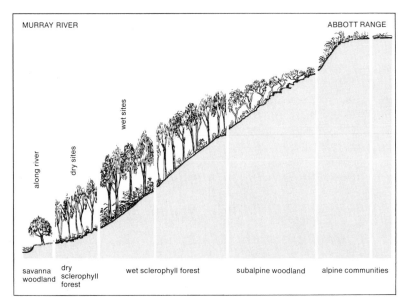

MURRAY RIVER   ABBOTT RANGE

along river | dry sites | wet sites

savanna | dry | wet sclerophyll forest | subalpine woodland | alpine communities
woodland | sclerophyll | | |
| forest | | |

Fig. 4 Generalized sequence of main plant communities from the Upper Murray
River to Abbott Range at Kosciusko. Each of the vegetation zones depicted comprises
a mosaic of communities, e.g. as shown in Figs 5 and 6, for the alpine zone.

sp.* and *Colobanthus nivicola†*, both of which are probably endemic to
the area. A few scattered herbs, notably the beautiful white Anemone
Buttercup (*Ranunculus anemoneus*), may also occur among the stones.
Because of the adaptation of feldmark species to cold bare ground, some
of them are also able to colonize areas of advanced soil erosion where
the topsoil has been removed down to a residual 'pavement' of stones
(Costin *et al.* 1959; Totterdell and Nebauer 1973).

On the very wind-exposed sites as found above and on the windward
side of the snowdrift areas another feldmark alliance develops, in this
instance characterized mainly by the prostrate shrubs *Epacris microphylla*,
E. petrophila and *Chionohebe densifolia*. In contrast with the *Coprosma–Colobanthus* feldmark, the *Epacris–Chionohebe* community receives
relatively little snow cover because the snow is blown off by the prevailing
winds, and the community is thus exposed to freezing conditions in winter
and spring when most of the alpine vegetation is protected by snow.
On some sites upwind erosion and downwind regeneration of clumps

*This is a new and as yet undescribed species of *Coprosma*, formerly thought to be *C.
pumila* to which it is related.
†This recently described species of *Colobanthus* and *C. pulvinatus* were formerly referred
to as *C. benthamianus*.

of this community are a natural process with the result that there is a slow migration of feldmark clumps in the direction of the prevailing westerly winds; the rate of movement has been measured at about 1 cm per year (Barrow *et al.* 1968). These exposed sites were also unfavourable during colder periods in the past when they were subjected to severe frost action, which in places penetrated to subsoil depths leading to downslope movement of wet surface soils over still-frozen subsoils (Costin *et al.* 1967). The soil terraces on the Main Range referred to earlier

General view of the Main Range from above Mt Twynam looking towards Mt Kosciusko capped with cloud shadow. The pale streaky patches below the large snow drift on the upper Blue Lake cirque are communities of *Neopaxia australasica*, a major component of short alpine herbfield, in flower. The extensive Mawson cirque is in the middle distance and Etheridge and Ramshead Ranges are to the left of Mt Kosciusko. The bare patch in the immediate foreground is sheet erosion, still active despite the cessation of grazing in the area over 30 years ago. More recent human pressures now threaten parts of this fragile area. The tracks and roads extending from the Mt Kosciusko summit to beyond Mt Twynam are situated mainly along the ridges occupied by the *Epacris–Chionohebe* feldmark, a community unique to Kosciusko and which takes up less than one-half of one percent of the alpine area.

were formed by these so-called solifluction processes, mostly during the recent cold phase 1500-3000 years ago. Where recent advanced erosion of nearby alpine herbfield areas has occurred, with the development of stony erosion pavements, some of the feldmark species, notably *Luzula oldfieldii* subsp. *dura*, are among the first colonizers.

Less rocky situations and areas of shallow stony well-drained soils sup-

TOP RIGHT   Clumps of *Epacris microphylla* on wind-exposed feldmark with a background of Sentinel Peak and the Townsend Spur. On the right is a small solifluction terrace with more continuous vegetation of tall alpine herbfield plants growing at its base.

BOTTOM RIGHT   A striking example of wind-induced pattern in *Epacris–Chionohebe* feldmark on the Main Range track south of Carruthers' Peak. Although the vegetation as a whole is stable, there is a slow movement of the individual clumps in a downwind direction owing to erosion on the exposed windward side and regeneration on the leeward side [253]. The stony pavement minimizes further wind erosion of the soil. These rare communities should be strictly preserved.

Solifluction terrace in *Epacris–Chionohebe* feldmark showing succession to tall alpine herbfield at the base of the terrace (see Fig. 6). The Townsend Spur is in the background.

port closed shrubberies, termed heaths, of the *Oxylobium–Podocarpus* alliance. The heaths in the more exposed sites (often grading into areas of *Epacris–Chionohebe* feldmark) are characterized by the dwarf myrtaceous shrub *Kunzea muelleri* which may form extensive even carpets up to 20 cm high. The *Kunzea* heaths in turn grade into taller heaths characterized by *Phebalium ovatifolium*, *Grevillea australis* and many other shrubs as well as *Oxylobium ellipticum* and *Podocarpus lawrencei* themselves. Glacial moraine and rocky solifluction deposits are typical habitats of this community. Where not destroyed by fire, some of the plants of *Podocarpus*, the dwarf Mountain Plum Pine, may reach several hundred years in age. *Podocarpus* is the longest-lived of all the alpine species at Kosciusko.

Another heath alliance, dominated mainly by *Epacris glacialis*, occurs under quite different conditions: damp, poorly drained soils surrounding wetter areas of bog and fen, with which it has been mapped for the purposes of this book.

The herbaceous communities, short and tall alpine herbfields and sod tussock grassland, together occupy the largest part of the alpine zone.

The most specialized alliance, short alpine herbfield characterized by *Plantago glacialis*, *Neopaxia australasica*, *Caltha introloba* and other dwarf mat-forming or creeping species, is found mainly below semi-permanent snow patches, near the *Coprosma–Colobanthus* feldmark. Here, during the short snow-free season, there is an abundance of snow-melt water which is relatively rich in nutrients owing to suspended debris derived from dust settling on the snow patch and rock debris eroded from beneath it. In some years, the rate of dust accumulation approaches 2000 kg/ha (Walker and Costin 1971). The short alpine herbfield vegetation itself is a peat-former, the oldest snow patch peats just below the large snow patches being apparently about 2500 years old. This indicates that the present snow patch regime has probably been in existence for about this length of time. Presumably, any older snow patch peats were eroded away by the increase in snow patch activity which would have occurred during the colder period 1500 to 3000 years ago referred to previously. Other situations with localized occurrence of short alpine herbfield are permanently wet gravelly and bed-rock surfaces such as occur along creeks, and also where there has been almost complete erosion of bog- or fen-peats.

The tall alpine herbfields comprise three main communities. The most widespread community is characterized by *Celmisia* sp. and species of *Poa* with many other conspicuous herbs; it occurs on most 'average' alpine sites, avoiding only areas of persistent snow cover, exposure to strong winds, and waterlogged or very stony soils. The tall, broad-leafed tussock

The steep western slopes of the Main Range looking south to the Townsend Spur from near Sentinel Peak. Heaths of the *Oxylobium–Podocarpus* alliance and tall alpine herbfield with *Chionochloa frigida* extend beyond the Snow Gum communities which form a very diffuse treeline owing to varying local conditions.

grass, *Chionochloa frigida*, characterizes a second community growing as a tall tussock grassland now found mainly on the steep, relatively inaccessible western slopes of Kosciusko, with much smaller patches on the more accessible east-facing slopes. This distribution pattern supports historical records that *Chionochloa* was formerly much more abundant (Costin

The south bank of Lake Albina consists largely of hummocky moraine left behind by previous glaciation in the valley. This area now supports extensive heath vegetation.

1958). Being palatable to livestock and very slow-growing, the grass became almost extinct during the earlier years of sheep- and cattle-grazing, except in inaccessible places. Since the elimination of grazing it has made a spectacular recovery. Many other alpine herbs have a similar history, including the Anemone Buttercup (*Ranunculus anemoneus*) and the Mountain Celery (*Aciphylla glacialis*), the latter often found with

Chionochloa. The third tall alpine herbfield community is restricted to rock faces, crevices and shady boulders. It is characterized mainly by *Brachycome nivalis* var. *nivalis* and *Danthonia alpicola*, with a number of ferns including *Polystichum proliferum, Blechnum penna-marina,* and the rare *Cystopteris fragilis* and *Grammitis armstrongii.*

With increasing soil wetness and restricted drainage – but before permanently swampy conditions develop – the tall alpine herbfield communities grade into sod tussock grassland, as in the broad valley of the Upper Snowy River. Stiff-leafed grasses – *Poa costiniana* and *Danthonia nudiflora* – are the most important dominants, with the smaller grass *Danthonia nivicola* and often with the early-flowering *Euphrasia collina* subsp. *glacialis* and the later-flowering Mountain Gentian (*Gentianella diemensis*). Because of difficulties in distinguishing some vegetation boundaries on the aerial photographs, sod tussock grassland is grouped with tall alpine herbfield on the vegetation map.

Permanently wet sites (other than short alpine herbfield below the snow patches) support bog or fen vegetation, often with the marginal wet heaths of *Epacris glacialis* referred to earlier. The bogs, characterized by the bog moss *Sphagnum cristatum*, are spring-fed by seepages down the hillsides or along the edges of the valleys. The hillside sites, although always wet, are better drained and are suitable for the development of a number of prickly-leafed shrubs which grow with the *Sphagnum* in the form of a slightly domed 'raised' bog, as the *Epacris–Sphagnum* alliance, with *Epacris paludosa* as the dominant shrub. The valley-edge sites are less suitable for the bog shrubs; here the *Sphagnum* is associated with sedges to form a flat or concave 'valley' bog of the *Carex–Sphagnum* alliance, with *Carex gaudichaudiana* as the dominant sedge. In still wetter parts of the valley, or wherever water ponds on relatively flat surfaces, the *Sphagnum* disappears and the sedges, notably *Carex gaudichaudiana*, form an acid fen. As noted previously, some of the fen peats are about 15 000 years old and give the best available estimates of when parts of the Kosciusko alpine area started to become free of glacier ice.

Some communities, especially the alpine herbfields on parts of the Main Range and the bogs and fens, have suffered from recent erosion and are not in their natural condition. Where the erosion has not been severe and some of the original topsoil remains, protection from grazing and fires has permitted recolonization by many native species, typically minor associates of the undisturbed community, and by various introduced herbs such as Sorrel which are adapted to bare soil. Where erosion has been more severe and has involved the formation of a stony erosion pavement, feldmark and short alpine herbfield species are among the colonizers, as has been noted already (cf. Totterdell and Nebauer 1973). Some of the most actively eroding areas, where natural recovery is too slow, have been re-vegetated with introduced grasses and clovers by the Soil Conservation Service of New South Wales (Bryant 1971).

The general distribution of the plant communities and other features of the alpine environment can be appreciated from the main vegetation maps. Table 5 shows the approximate areas occupied by these communities and features, and their relative abundance. It will be seen that tall alpine herbfields and associated sod tussock grassland cover more than half of the alpine area, and heaths about one-quarter; by contrast, the two feldmark communities and short alpine herbfield are the most restricted types of vegetation, together occupying less than 3% of the area.

Although the maps are at a relatively large scale they still over-simplify

TABLE 5 Areas of main plant communities and associated features mapped in the Kosciusko alpine zone

Mapping unit	Area: (sq miles)	(ha)	Percentage of mapped area
Feldmark			
Epacris–Chionohebe	0·11	28·5	0·30
Coprosma–Colobanthus (and associated snow patch areas)	0·60	156·0	1·62
Short alpine herbfield			
Plantago–Neopaxia	0·32	82·2	0·85
Tall alpine herbfield			
Celmisia–Poa [A, B]	20·62	5340·6	55·34
Chionochloa	2·34	606·7	6·29
Heath			
Oxylobium–Podocarpus	8·43	2182·1	22·61
Kunzea	1·03	266·8	2·76
Fen and bogs			
Carex, Sphagnum and associated epacrids[C] etc.	2·16	559·4	5·80
Subalpine woodland			
Eucalyptus pauciflora subsp. *niphophila*	1·13	292·0	3·03
Lakes	0·11	27·8	0·29
Rock screes	0·16	41·4	0·43
Rock outcrops [A]	0·21	53·1	0·55
Eroded areas	0·05	13·6	0·14
Total alpine area (and associated islands of subalpine woodland)	37·26	9650·3	

[A] Also includes small areas of *Brachycome–Danthonia* tall alpine herbfield.
[B] Also includes areas of *Poa–Danthonia* sod tussock grassland.
[C] Also includes associated areas of *Epacris glacialis* heath.

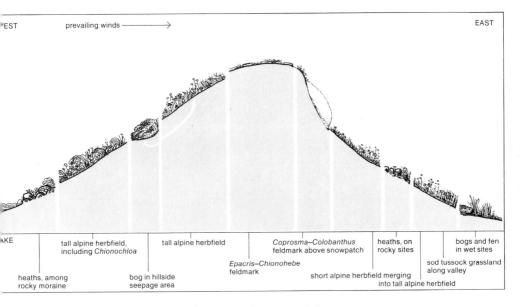

Fig. 5 The alpine vegetation at Kosciusko consists of a mosaic of plant communities, distributed according to variations in site conditions, as shown in this diagrammatic section of part of the Main Range.

much of the spatial variation in the vegetation which actually occurs. In response to the abrupt gradients in environmental conditions often found in the one small area of the alpine zone, the vegetation also varies from place to place, often in a fine mosaic pattern. A typical example of the variation is shown in Fig. 5, illustrating the transect from a wind- and sun-exposed west-facing slope across the ridge to the leeward shady side. On a still finer scale, the variation over a series of exposed solifluction terraces is illustrated in Fig. 6a.

The interpretation of spatial variations in plant distribution is one of the main problems of plant ecology. What are the causes of such variations in any given area, and what effects progressively do the plants themselves have on what initially may have been critical environmental factors? This leads to questions such as: have existing patterns been different in the past, and will they change in future? There is no general answer to such questions and much depends on the local situation. Thus some of the patterns, as illustrated in Fig. 5, are relatively permanent (hundreds to thousands of years) except in the event of a considerable climatic change, especially in temperature. But others, of the type illustrated in Fig. 6a, may represent relatively rapid changes in plant distribution with time

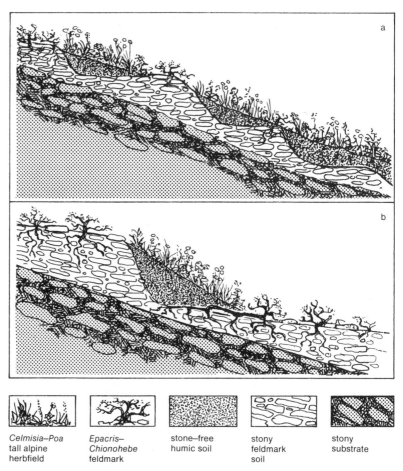

Celmisia–Poa
tall alpine
herbfield

Epacris–
Chionohebe
feldmark

stone–free
humic soil

stony
feldmark
soil

stony
substrate

Fig. 6a Cross section through a downslope sequence of solifluction terraces along the Main Range showing variation in the composition of the terraces with depth, and the associated pattern of feldmark and herbfield vegetation.

Fig. 6b Cross section through adjacent solifluction terraces showing, in addition to features described in Fig 6a, the succession of feldmark by herbfield vegetation as the base of the lower terrace and associated *Epacris* plants becomes covered with the fine soil eroded from the surface of the upper terrace.

as well as in space. Studies on the Mt Clarke and Mt Lee areas show that although the condition illustrated in Fig. 6a is relatively permanent in many sites, it may be a developmental one in others, in a succession from *Epacris–Chionohebe* feldmark to *Celmisia–Poa* herbfield, as shown in Fig. 6b. Fig. 6b indicates how fine soil from the sparsely vegetated feldmark surface of the upper terrace is deposited at the base of the adjacent

terrace, gradually covering the *Epacris* vegetation there as it provides more
suitable conditions for the growth of herbfield species. Gradually, many
feldmark terraces may thus become completely covered by herbfield vegetation and soils, to the extent that the original stepped terrain is smoothed
over and virtually obscured. However, soil borings along downslope tran-
sects will often reveal a regular variation in depth of the surface soil above

Aerial view of extensive contour terracing on gently sloping terrain near Merritt's
Creek [50].

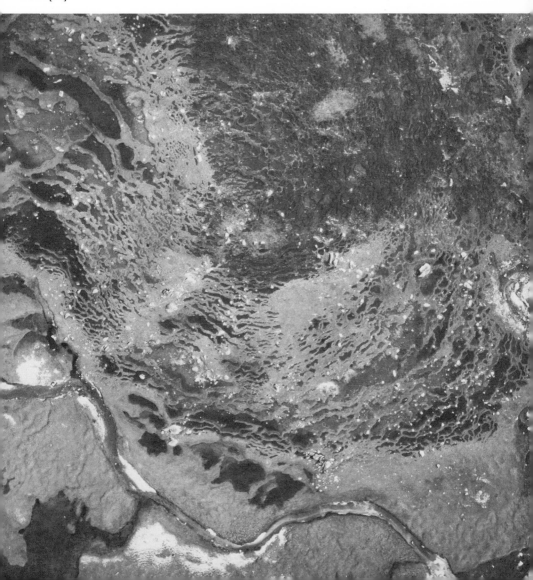

the now buried feldmark surface, as illustrated in Fig. 6a. Such processes of plant succession and soil development were apparently important immediately following earlier colder periods, leading to rapid stabilization of the landscape. The above-mentioned studies indicate that such stabilization can occur locally in as short a period as 50 years (Costin *et al.* 1969).

Similarly there are swampy areas where there has been a temporal sequence in the plant cover corresponding to the spatial sequence seen from the middle to the edge of a pool – in this instance from *Carex* fen to *Epacris–Sphagnum* bog – within a period of several hundred years. Yet nearby, other swampy sites apparently have been occupied by fen, without change to bog, for many thousands of years.

Natural vegetation may also experience periodically severe stresses that induce local changes and consequent variety in the plant communities. In tall alpine herbfields and sod tussock grassland, for example, patches of snow grasses are sometimes damaged by insects, prolonged snow cover and drought. Some of these areas are re-colonized by broad-leafed herbs (especially Compositae), and other stonier sites by shrubs, which may remain locally dominant for a few years or up to several decades while the snow grasses slowly re-establish themselves. Similarly some heath shrubs, particularly where growing on exposed shallow soils (as *Kunzea muelleri* often does), may be killed by the occasional drought and replaced by herbs. Successional changes such as these are normal within many types of vegetation and enhance the effects of site differences in giving the vegetation complexity and variety. A policy of strict preservation, as is most appropriate in the Kosciusko alpine area, then becomes one of the most positive and powerful management tools at the ecologist's disposal since it accepts most natural disturbances and the resultant plant successions. Proponents of more 'active management' for national parks and nature reserves – suitable for some types of vegetation and various species of wildlife – should first evaluate their expected results against the often far more complex and diversifying consequences of the so-called 'do-nothing' policy of preservation.

The naturalist at Kosciusko – whether amateur or professional – will find himself confronted by many intriguing problems of plant distribution

34 *Podocarpus lawrencei*, one of the characteristic components of the *Oxylobium–Podocarpus* heath alliance, is seen here in typical rock-sprawling habit. The main roots of the trunk are in the soil near the base of the rocks, but secondary roots develop from the branches and permeate cracks and crevices on the rock faces.
35 Heath of the *Oxylobium–Podocarpus* alliance typically prefers rocky sites such as this moraine above Lake Albina.

34

35

36 37

36 The heaths vary greatly in floristic composition and in structure. This photograph shows *Phebalium ovatifolium*, with its characteristic flat-topped crown, forming a mixed community with tall alpine herbfield. The bare cliffs of the Mt Clarke cirque are seen on the left.

37 *Kunzea muelleri*, seen here in flower, forms extensive low heaths in exposed rocky sites. Under less severe conditions it grades into heaths of taller-growing species.
38 Damp sites marginal to bogs, lakes and streams etc. support low-growing heaths of the *Epacris glacialis* alliance, seen here as purplish brown patches near Lake Cootapatamba.

39 Tall alpine herbfield, as seen in the foreground, includes a wide variety of herbaceous species and is the most widespread community covering the rounded hills of the Main Range. The white-flowered, silvery-leaved daisy is *Celmisia*, the mauve flowers *Euphrasia* and a few yellow flowers of *Craspedia* are also visible.

40 Extensive and continuous communities of flowering Silver Snow Daisy (*Celmisia* sp.), a dominant herb of the *Celmisia–Poa* tall alpine herbfield alliance, can be seen throughout the Kosciusko alpine area during January and early February. Mt Townsend is in the background.

41 Typical steep rocky habitat of *Brachycome–Danthonia* tall alpine herbfield alliance on the cliffs above Blue Lake showing the *Brachycome* in flower. The rocky sites below, where long-lasting snow drifts persist, support small bright green patches of short alpine herbfield.

42 Many bare and semi-bare areas at Kosciusko are being colonized by the naturalized species *Rumex acetosella* (Sheep Sorrel), which forms bright reddish patches when in flower. Mt Townsend and Mueller's Peak are in the background.

43 Short alpine herbfield of the *Plantago–Neopaxia* alliance typically occurs in damp sites below snow patches irrigated by snow-melt waters containing nutrients in the form of wind-blown dust and other debris which collects on the snow surface [29]. Here *Neopaxia australasica* is seen in flower with *Plantago glacialis* and a few tufts of *Carex gaudichaudiana*. The stony snow patch area is in the background.

42

43

44 Short alpine herbfield growing along melt-water streams below Carruthers' Peak. *Caltha introloba* is in flower in the foreground.
45 Prolific flowering of *Ranunculus niphophilus* in short alpine herbfield at the head of Carruthers' Creek. *Carex* fen surrounds the stream in the gully.
46 Slopes irrigated by snow-melt streams are favoured habitats of short alpine herbfield plants.

45

46

47 Low-lying area east of Seaman's Hut, showing
sod tussock grassland with red-stemmed *Danthonia
nivicola* in background and straw-coloured stems of
Danthonia nudiflora in foreground. The green sward
is mainly composed of *Poa* grasses. Mt Twynam and
Little Twynam can be seen in the background.

48 Upper Snowy River Valley with extensive sod
tussock grassland in which the red-stemmed grass
Danthonia nivicola is a characteristic species. Note
also the small dark green fen community of *Carex
gaudichaudiana* in wet sites along the stream.

49 *Poa* tussocks in sod tussock grassland.

50 In even, gently sloping situations contour terraces
occur in which the sod tussock grassland is mainly
confined to the banks of the terraces, with characteristic pool species such as *Ranunculus millanii*, *Myriophyllum pedunculatum*, *Pratia surrepens* and *Agrostis
meionectes* in the depressions. The contour terraces
may be solifluction features formed during colder
conditions of deep seasonal soil freezing and thawing.

48

47

49

51 Spring growth of *Carex gaudichaudiana* in snow-melt water near Hedley Tarn. Some of the peats associated with these *Carex* fens are up to 15 000 years old, indicating that these parts of the alpine landscape were then emerging from the cover of glacial ice and snow of the last main Ice Age.

52 A high-altitude fen of the *Carex gaudichaudiana* alliance near Carruthers' Peak showing the gradation of the dark green fen into short alpine herbfield beneath the snow patch area. Tall alpine herbfield with a patch of *Celmisia* is on the left. The apparently bare snow patch area supports feldmark vegetation of the *Coprosma–Colobanthus* alliance.

53 Wet basin above Lake Albina showing *Carex* gaudichaudiana fen with small patches of yellowish *Sphagnum cristatum* moss along the stream.

54 Along the better-drained edges of the valley and around hillside springs raised bogs of the *Epacris–Sphagnum* alliance develop, with many species of shrubs as well as the *Sphagnum* and *Carex. Epacris paludosa*, in flower near Club Lake, and *Richea continentis* are typical components of raised bogs.

54

52

53

55 Valley bog with hummocks and banks of *Sphagnum cristatum* around pools with *Carex gaudichaudiana*.
56 *Epacris glacialis* (in flower) and *Empodisma minus* are found in *Sphagnum* hummocks of valley bogs.
57 Late summer view of part of the Main Range with Mt Kosciusko on the far left. The bare leeward east-facing slopes are free of deep snow for only a few months of the year and support sparse vegetation of *Coprosma– Colobanthus* feldmark. Directly below these is *Plantago–Neopaxia* short alpine herbfield. On the plateau the most wind-exposed ridges and slopes support *Epacris–Chionohebe* feldmarks which alternate with *Celmisia–Poa* tall alpine herbfield in more sheltered sites. The darker green vegetation is mainly heath of the *Oxylobium–Podocarpus* alliance.

56

57

58 Although most of the Kosciusko alpine plant communities are relatively stable as a whole, some undergo cyclical fluctuations as in *Epacris–Chionohebe* feldmark and in some alpine herbfield communities such as those shown here on the steep slopes above the upper Blue Lake cirque. The extensive light brown areas are snow grasses which died following the 1967–68 drought. However, some of the associated herbs survived and have since spread vigorously in the shelter of the mulch of dead snow grass leaves. The result now is a community relatively richer in herbs than the original snow grass community. In time, the snow grasses will tend to reassert themselves but may again succumb to natural stresses such as drought, prolonged snow cover or attack by insects. Such cyclical changes are not uncommon in plant and animal communities.

59 Throughout the growing season the alpine vegetation is subject to a wide range of climatic changes. An early January snowfall covers sod tussock grassland near Merritt's Creek.

60 Formerly regarded mainly as a place for winter recreation, the Kosciusko alpine area is now becoming increasingly popular in summer for nature-lovers and walkers, seen here approaching the alpine area through heath at the edge of the treeline.

in relation to associated environmental conditions and in relation to past and present changes in these conditions. His ability to solve such problems will be in proportion to his knowledge of the alpine plants and their habitats.

THE VEGETATION MAPS   The colour maps on pp. 13-15 show the distribution of the main types of vegetation and associated features in the Kosciusko Primitive Area (Wimbush and Costin 1973). The Primitive Area, established in 1963, extends from the Geehi River in the north-west and the Alpine Way in the west and south to the Crackenback Range in the south-east and the ridges between Mt Guthrie and Mt Tate in the north-east.

The line map, on p. 108, shows the altitudinal sequence of the major types of vegetation from the woodlands on the flats near Geehi, through dry and wet sclerophyll forests on the steep montane slopes of the Main Range, to subalpine woodland and alpine vegetation at higher levels. This is a characteristic vegetation sequence in many high mountain areas of Australia.

The colour maps delineate the main plant communities and associated features in the Kosciusko alpine area itself together with small areas of subalpine woodland along the north-eastern side. The names of the characteristic species of the alliances shown on these maps differ in part from those used in earlier publications (e.g. Costin 1954, 1961; Wimbush and Costin 1973). However, no confusion should arise, as the synonymy is indicated in the footnote to Table 4.

We anticipate that the users of this book will make frequent reference to these maps, not only for the locality and topographic information they contain but especially as a means of locating particular types of vegetation and the main species of plants which occur there.

However, it should be noted that, although drawn at a relatively large scale, the colour maps still simplify the distribution of the plant communities. Most areas mapped as a single plant community also contain one or more minor communities (as regards area) as well as the major (most extensive) one. Thus, most areas of *Chionochloa* tall alpine herbfield on the western slopes of the Main Range also include some *Oxylobium–Podocarpus* heath and vice versa. Similarly, some areas of *Celmisia–Poa* tall alpine herbfield include areas of *Poa–Danthonia* sod tussock grassland, areas of bogs and fen are commonly surrounded by *Epacris glacialis* heath, and areas of *Epacris–Chionohebe* feldmark usually contain patches of tall alpine herbfields and heaths. Furthermore, where it has not been possible directly to identify a plant community on the aerial photographs, but only associated environmental features, some over- or under-estimate

The Snowy River leaves the alpine zone below
Charlotte's Pass where Snow Gums form the treeline.
The distant skyline features are Etheridge Range,
Mt Kosciusko and Mts Northcote and Clarke.

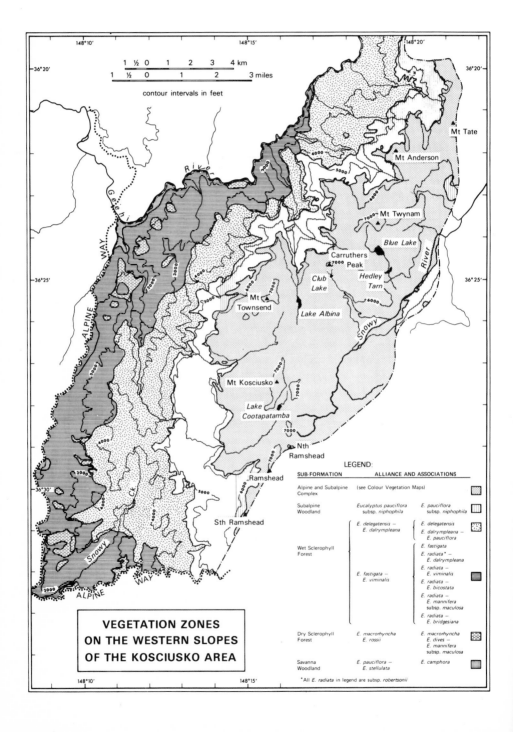

LEGEND:

SUB-FORMATION	ALLIANCE AND ASSOCIATIONS		
Alpine and Subalpine Complex	(see Colour Vegetation Maps)		
Subalpine Woodland	*Eucalyptus pauciflora* subsp. *niphophila*	*E. pauciflora* subsp. *niphophila*	
Wet Sclerophyll Forest	*E. delegatensis* — *E. dalrympleana*	*E. delegatensis*	
		E. dalrympleana — *E. pauciflora*	
	E. fastigata — *E. viminalis*	*E. fastigata*	
		*E. radiata** — *E. dalrympleana*	
		E. radiata — *E. viminalis*	
		E. radiata — *E. bicostata*	
		E. radiata — *E. mannifera* subsp. *maculosa*	
		E. radiata — *E. bridgesiana*	
Dry Sclerophyll Forest	*E. macrorhyncha* *E. rossii*	*E. macrorhyncha* — *E. dives* — *E. mannifera* subsp. *maculosa*	
Savanna Woodland	*E. pauciflora* — *E. stellulata*	*E. camphora*	

*All *E. radiata* in legend are subsp. *robertsonii*

VEGETATION ZONES ON THE WESTERN SLOPES OF THE KOSCIUSKO AREA

of area may have resulted. This is the case with *Coprosma–Colobanthus* feldmark which has been mapped in terms of the large snow patch areas with which it is typically associated and has probably been over-estimated in extent as a result.

However, these problems are inherent in most vegetation maps. Because of its large scale, the map of the Kosciusko alpine area is more realistic than most in reflecting the actual extent and distribution of the main plant communities in relation to important environmental factors (Table 4).

IDENTIFICATION AND DESCRIPTION We have seen that plants can be classified according to several criteria, including growth form, life form and geographic affinities, to name only a few. But the most widely used classification of plants is the taxonomic classification based on the properties of the flowers and fruits, with subsidiary emphasis on vegetative characteristics, growth habit etc. This classification and the binomial system of nomenclature, which were first consistently applied by the Swedish botanist Linnaeus, provide a universal language among botanists.

The identification of plants according to their taxonomic characteristics requires some knowledge of structural botany. The use of technical language is unavoidable and frequent consultation of the glossary may be necessary for those not familiar with botanical terms. However, those who take the initial trouble to key out the plants rather than take the short cut of matching them against the illustrations will find, to their satisfaction, that they are also training themselves to identify plants in any environment. A hand-lens, with about a 10× magnification, is essential for the close examination of many species.

Keys to the Kosciusko alpine flora and descriptions of the individual plants with notes on distribution and habitat now follow; plate references are indicated by the numbers in square brackets which appear in the margins. Naturalized species, although included in the keys, have not been described and are indicated throughout by an asterisk. Although synonyms have been included in the index, the synonymy is not exhaustive and is restricted to basionyms and to those names most likely to be encountered in the older literature on the Kosciusko flora. We have also included some common names mainly following Willis (1970, 1973), but for many species there are as yet no accepted common names and we have not attempted to introduce any except for a few which seem appropriate. This is because common names, unlike scientific names, tend to evolve as the species in question become familiar to people.

Names of colours used in the descriptions generally follow the *Methuen Handbook of Colour* (1963).

As noted earlier, we are confining our attention to the strictly alpine species of vascular plants at Kosciusko, and for the purposes of this book we have arbitrarily delimited an area with the boundary at about 1830 m (6000 ft) or the treeline, whichever is the higher; this boundary is shown on the colour maps. Thus we exclude some marginally alpine species which occur amongst the isolated clumps of Snow Gum shown in the north-east corner of the colour map. Nevertheless, it will be found that many of the species treated also occur throughout the snow country of mainland Australia and Tasmania, a considerable number also grow at lower levels, and some are found in other countries.

It will be seen that our understanding of the flora is still far from complete and that some of the quite common and important genera either require revision or are undergoing revision at the present time. Rather than postpone publication of the *Flora* for several years until most of the outstanding problems could be solved, we have in some cases indicated partial solutions, and otherwise have done our best to point out where the deficiencies lie. No doubt this will offend the perfectionists. However, we have taken heart from the following thought expressed by V. H. Heywood (1971): 'We adopt the view that an honestly produced Flora, no matter what its defects are, can be revised. It is much more difficult to revise a non-existent Flora!'

The Alpine Flora

The keys here have been much abbreviated, mainly because they have been designed for use in the field but also because many of the larger groups of plants are only represented by one or a few species. The main aim has been to lead the user to the species as soon as possible rather than by logically progressing through the hierarchy of taxa in an orthodox fashion; this has resulted in many short cuts which would not be possible or desirable in a larger flora. Apart from this, the keys are of the ordinary dichotomous kind, that is, they are composed of pairs of sets of contrasting characters, each alternative pair with the same number. The characters in numbers 1 and 1 are first compared, and the alternative which best fits the specimen to be identified is chosen. This leads either to a name or to another number. In the latter case, proceed down the key directly to the number indicated and again compare the contrasting characters in that number pair and choose the correct alternative. This again leads either to a name or to another number and in the latter case the same process is repeated until the name of the plant is arrived at.

---

1 Ferns and related plants, without flowers or seeds, reproducing by spores GENERAL KEY
   PTERIDOPHYTA (p. 112)
1 Not as above; plants reproducing by seeds 2

---

2 Leafy, much-branched shrub, prostrate and creeping over rocks and up to several metres in diam.; perianth absent; pollen sacs on microsporophylls arranged in small cylindrical cones 5–10(–25) × 1.5 mm; seed nut-like, naked on a receptacle of 2 fused fleshy scales which become swollen, subglobular, red and succulent at maturity. (The seed of *Exocarpos nanus* on its swollen red fleshy peduncle is similar to this in appearance; however, the latter is a much smaller plant with flattened branchlets and leaves reduced to scales)
   Podocarpus lawrencei (GYMNOSPERMAE) (p. 118)
2 Not as above; perianth present but sometimes modified or inconspicuous; pollen sacs in anthers of stamens; ovules and seeds enclosed within carpels which bear stigmas on which the pollen lodges (ANGIOSPERMAE) 3

---

3 Flower parts usually in 3s or multiples of 3; if the perianth absent or reduced to scales or bristles then the flowers ± hidden in the axils of glumaceous bracts (grasses, sedges, rushes, orchids and lilies)
   Monocotyledoneae (p. 119)
3 Flower parts usually in 2s, 4s or 5s or multiples thereof; if the perianth reduced or inconspicuous then the flowers not hidden in the axils of glumaceous bracts Dicotyledoneae (p. 161)

PTERIDOPHYTA

Lycopodiaceae

Lycopodium L.

L. fastigiatum R. Br., Prodr. Flor. Nov. Holl. 165 (1810)
'Mountain clubmoss'

Pale or yellowish green glabrous perennial; *main stem* rhizome-like, long-creeping [61] and rooting, sparsely scaly, about 1.5 mm diam.; *aerial branches* lax, much-branched, erect, trailing or decumbent up to 15 cm or more high, overtopped by the fertile spikes; *leaves* crowded, ± 3–4 × 0.7–1 mm, linear, acute, curving upwards; *fertile spikes* 2–5 cm long, sometimes forked, solitary or few together on long peduncles which bear small sparse bract-like leaves; *sporophylls* peltate, broader than the leaves, 3–4 × 1–1.5 mm, ovate, acuminate, the margins scarious and minutely lacerate proximally, the tips usually spreading or recurved at maturity; *sporangia* yellow, 1–1.3 × 1.5–1.9 mm, broadly reniform, opening by a transverse slit.

Mountainous areas of south-eastern Queensland, eastern N.S.W., Victoria and *Distribution* Tasmania; also New Zealand (North and South Islands) and Stewart, Chatham, Auckland, Campbell and Antipodes Islands.

This species is sparsely distributed, but not uncommon, in the alpine and subalpine *Notes and* tracts. Typical habitats are the bases of rocks, under shrubs, around bogs, and *Habitat* it is not uncommon in tall alpine herbfields and on bare patches of soil.

References: Beadle (1971); Tindale (1955a).

Huperzia Bernh.

H. selago (L.) Bernh. ex Schrank & Mart., Hort. Monac. 3 (1829)
'Fir clubmoss'

Dark green glabrous perennial to ± 15 cm high; *stems* dichotomously branched, [62] ascending or suberect, shortly creeping and rooting at the base; *leaves* crowded, 5–7.5 × 1–1.6 mm, linear-lanceolate, acute, entire or serrulate distally, sometimes with small reproductive buds in the axils; *sporangia* yellow, 0.8–1.1 × 1–1.5 mm, broadly reniform, opening by a transverse slit, sessile and solitary in the axils of leaf-like sporophylls.

Sparsely distributed in the alpine and subalpine tracts of the Australian Alps, *Distribution* and mountains of Tasmania and New Zealand (see Notes below).

The *H. selago* complex is widespread in cool montane areas throughout the world, *Notes and* and the status of the Australasian representative needs further investigation *Habitat* (Allan 1961; Willis 1970). *H. selago* is fairly rare above treeline, but occurs on the shores of Blue Lake and on the slopes of Mt Stilwell. It usually grows in dense colonies in moist shady places, often under large boulders, and in bogs.

References: Allan (1961); Willis (1970).

Grammitidaceae

Grammitis Sw.

G. armstrongii Tindale, Contrib. N.S.W. Natl Herb. 3: 88 (1961)

[63]   Dwarf fern, usually growing in mats or patches among moss in crevices of boulders, rock-faces etc.; *rhizomes* thin, branched, long-creeping, densely covered with thin yellowish brown or pale ferruginous scales; *leaves* coriaceous to herbaceous, crowded, subglabrous, 1–3.5 cm × 2–5 mm, oblanceolate, narrowly oblanceolate or spathulate, obtuse or rounded at the apex, gradually narrowed to the petiole, the margins entire or obscurely sinuate distally; *sori* without indusia, becoming confluent in a rounded mass in the distal portion of the lamina at maturity.

Distribution   Alpine and subalpine tracts of the Australian Alps and summits of high mountains in Tasmania; also mountains of New Zealand (North and South Islands), Stewart Island, Macquarie Island and Kerguelen Island.*

Notes and   This species is not common above treeline but it occurs in *Brachycome–Danthonia*
Habitat   tall alpine herbfield on exposed rock faces and large boulders near Lake Albina, Blue Lake, Club Lake and on the south-east slopes of Mt Clarke. It is closely related to *G. billardieri* Willd., which occurs at lower altitudes, but the latter differs in having a tufted or shortly creeping rhizome, smaller spores, and the oblong sori are obliquely arranged on either side of the midrib and are seldom confluent at maturity.

References: Parris (1975); Tindale (1961).

*Parris (1975) has recorded a circumantarctic distribution for this species extending the above distribution to include: Campbell and Auckland Islands, Chile, Argentina, Falkland Island, South Georgia, Tristan da Cunha, Gough Island, South Africa (Cape Province) and Marion and Crozet Islands.

Aspleniaceae

Asplenium L.

A. flabellifolium Cav., Descr. Plant. 257 (1802)   'Necklace fern'

[64]   Delicate fern with lax, trailing stems, the short rhizome and the base of the petiole covered with small, dark, clathrate, narrowly triangular, attenuate scales ± 3–4 mm long; *leaves* 1–2(–2.5) cm wide, once-pinnate, the lower pinnae

sometimes ± lobed, sparsely covered with weak microscopic septate-glandular hairs, the petiole and rachis with occasional dark crinkled hair-like scales; *rachis* usually produced into a long naked tendril-like tip which may be proliferous; *pinnae* fan-shaped or wedge-shaped, very shortly petiolate, crenate or coarsely toothed or lobed distally, the lowermost up to ± 1 cm long and sometimes subopposite, becoming alternate and decreasing in size up the rachis; *sori* linear or oblong, obliquely arranged 2–5(–8) per pinna; *indusium* pale, membranous, opening along the inner margin, finally ± hidden by the dense mass of sporangia which may cover the underside of the pinnae.

Widespread in temperate Australia including Tasmania and in New Zealand. *Distribution*

Although common at lower elevations, Necklace Fern is rare above the treeline, being known from near Blue Lake and Seaman's Hut. Typically, it grows on sunny aspects under rocks and in crevices as a minor component of tall alpine herbfield.

Notes and Habitat

Blechnaceae

Blechnum L.

B. penna-marina (Poir.) Kuhn, Fil. Afr. 92 (1868)

'Alpine water-fern'

Small fern 5–20(–35) cm high; *rhizomes* wiry, long-creeping, 2–3 mm diam., with sparse pale brown or ferruginous papery scales; *leaves* subcoriaceous, pinnatisect, the fertile and sterile leaves dissimilar; *sterile leaves*: erect or ascending, about 5–15 cm long, with ovate or ovate-lanceolate shiny brown scales at the base of the petiole; *lamina* ± 1–2 cm wide, narrowly elliptical in outline; *leaf-segments* close-set, alternate or opposite, attached by their full width, dark green and conspicuously nerved above, paler on the underside, the margins slightly recurved; *fertile leaves*: erect, at length overtopping the sterile leaves, usually with a few reduced sterile leaf-segments at the base of the lamina; *leaf-segments* distant, narrow-oblong, ± arcuate; *indusium* opening along the inside, the sori confluent at maturity, the red-brown mass of sporangia covering the underside of the pinnae.

[65]

Great Dividing Range from Tenterfield, N.S.W., to Victoria, and mountains of Tasmania; also New Zealand, temperate South America, and widespread in the subantarctic islands.

Distribution

This is a common and widespread species above the treeline, found along streams, in tall alpine herbfields and in other communities.

Notes and Habitat

Reference: Beadle (1971).

Aspidiaceae

Polystichum Roth

P. proliferum (R. Br.) Presl, Tent. Pterid. 83 (1836)
'Mother shield-fern'

[66] Coarse tufted fern about 20–60 cm high (to 100 cm at lower altitudes), usually growing in dense masses along crevices between rock outcrops; *rhizomes* short, erect or ascending, thick and woody, densely covered with dark or ferruginous, linear-attenuate, spirally twisted scales; *leaves* coriaceous, the primary rachis usually with a few small dark scaly (occasionally proliferous) buds towards the apex; *laminae* 2(–3)-pinnate, dark green above, paler on the underside, the rachises with fine, soft, ferruginous ± twisted hair-like scales; *pinnules* asymmetrical; *petioles* densely covered towards the base with coarse ± falcate dark red-brown or black, usually pale-bordered, triangular-acuminate glossy scales; *sori* orbicular, often confluent at maturity; *indusium* orbicular-peltate, centrally attached, ± 1 mm diam., pale with a small dark central spot.

Distribution Tablelands and ranges of eastern New South Wales, Victoria and Tasmania.

Notes and Habitat This fern is common in sheltered sites around the bases of boulders and rock outcrops, as an important component of *Brachycome–Danthonia* tall alpine herbfield; also in other tall alpine herbfields.

References: Tindale (1955*b*, 1961).

Athyriaceae

Cystopteris Bernh.

C. fragilis (L.) Bernh., Schrad. Neues J. Bot. 1(2): 27 (1805)
'Brittle bladder-fern'

[67] Delicate tufted fern about 10–30 cm high; *rhizomes* short, ascending or shortly creeping, with soft, crinkled, attenuate, pale brown clathrate scales; *leaves* variably dissected, narrow-lanceolate in outline, once-pinnate, or 2-pinnate towards the base; *pinnae* distant, subopposite or alternate, the ultimate segments obtuse or rounded; *petioles* with brown crinkled clathrate scales towards the base, distally pale stramineous and appearing glabrous, but usually with sparse microscopic glandular-septate hairs which extend to the rachises; *sori* orbicular; *indusia* pale, membranous, hood-shaped, attached at the base and free at the pointed apex.

Alpine and subalpine tracts of the Australian Alps and mountains of Tasmania; *Distribution*
the species, in the broad sense, is cosmopolitan in mountainous areas of the world.

The Australian form of this widespread species has been referred to var. *laetevirens* *Notes and*
(Prent.) C. Christ. It is very rare above the treeline, in wet shady rock crevices *Habitat*
in crags above Blue Lake, as a minor component of *Brachycome–Danthonia* tall
alpine herbfield.

GYMNOSPERMAE

Podocarpaceae

Podocarpus L'Hér. ex Pers.

P. lawrencei Hook. f., Hook. Lond. J. Bot. 4: 151 (1845)

'Mountain plum pine'

[68–70] Glabrous aromatic much-branched decumbent shrub, ± 0.5-1 m high and often several metres in diam., sprawling and creeping extensively over large rock outcrops, the young shoots reddish purple; *leaves* crowded, spreading, obscurely 2-ranked, coriaceous, ± 5–15 × 1.5–2.5 mm, linear-oblong to narrowly obovate, obtuse or rounded and sometimes minutely apiculate at the apex, dark green above, paler and with a prominent midrib on the underside; *male cones* cylindrical, 5–10(–15) × 1.5 mm, reddish purple becoming yellowish with age, solitary or 2–6 together on peduncles 3–5(–10) mm long; *female cones* sessile or on very short peduncles up to 3 mm long; *receptacle* of 2 fused fleshy scales with free tips, at first pale and ± 3–5 mm long, expanding at maturity to become bright red, succulent, subglobular and ± 7 × 6 mm; *seed* green, nut-like, naked on the swollen receptacle, ± 3–6 × 3–4 mm, ovoid and crested.

Distribution Australian Alps and associated ranges from the Brindabella Range, A.C.T., to Lake Mountain and the Baw Baws in Victoria, and high mountains of Tasmania.

Notes and Habitat This species of pine, the only alpine conifer on the mainland, is an important dominant of *Oxylobium–Podocarpus* heath; typically it occurs on rocky sites, including rock faces, screes and moraines. It was probably a pioneer and colonizing species of such sites following earlier glacial and peri-glacial conditions, because of its ability to form prostrate spreading mats from which new roots are developed, its capacity to fix atmospheric nitrogen by means of root nodules (Bergersen and Costin 1964) and its longevity. Where not destroyed by fire, to which it is very sensitive, *Podocarpus* may attain ages of several hundred years. Under alpine conditions it has a slow rate of growth, with trunk-diameter increments of as little as 0.25 mm per annum, but makes better progress under cultivation in parks and gardens. In sheltered sites at lower elevations, e.g. in Gippsland, Victoria, it grows as an erect small tree to ± 7 m high.

References: Bergersen and Costin (1964); Gray, N. E. (1956).

ANGIOSPERMAE: Monocotyledoneae

KEY

1 Perianth absent or reduced to scales or bristles and the flowers ± hidden in the axils of glumaceous bracts, or perianth glumaceous (grasses, sedges and rushes)   2
1 Perianth (at least the inner series) ± petaloid (orchids and lilies)   5

2 Perianth of 6 subequal glumaceous tepals; fruit a capsule
                                                            JUNCACEAE (p. 150)
2 Perianth absent or reduced to small scales or bristles, the flowers enclosed in glumaceous bracts; fruit a nut or caryopsis   3

3 Dioecious plants with wiry stems; leaves with rudimentary laminae 1–4 mm long, otherwise reduced to sheathing bracts up the stem
                                                   Empodisma minus (p. 149)
3 Not as above   4

4 Flowers distichously arranged with each flower enclosed in a bract and bracteole (lemma and palea), or if in 1-flowered spikelets then the perianth of microscopic scales (lodicules – not visible to the naked eye); anthers dorsifixed   GRAMINEAE (p. 119)
4 Flowers arranged in cylindrical spikes, with each flower (sometimes enclosed in a utricle) in the axis of a single bract (glume), or if in 1-flowered spikelets then the perianth of hypogynous bristles or scales which are half as long to longer than the nut; anthers basifixed   CYPERACEAE (p. 136)

5 Ovary inferior; perianth irregular   ORCHIDACEAE (p. 158)
5 Ovary superior; perianth regular   LILIACEAE (p. 156)

Gramineae

KEY

1 Spikelets all sessile and arranged in a simple spike, or spikelets in groups of 3 which fall together at maturity   2
1 Spikelets not as above   4

2 Spikelets in groups of 3 which fall together at maturity; glumes and lemmas
   with long awns 1.5–2.5 cm long; annual to 30 cm high with dense cylindrical
   spike-like inflorescence which readily breaks up at maturity; locally common
   on Soil Conservation reclamation areas   Barley Grass *Hordeum glaucum*
2 Spikelets solitary at each node of the rachis; glumes and lemmas unawned
   or with short awns up to ± 3 mm long   3

3 Spike long and slender, the spikelets distichous with their narrow edges fitting
   into hollows in the rachis; glume single and external except for the terminal
   spikelet which has 2 glumes; lemma awnless; slender perennial to about 45 cm
   high; locally common on Soil Conservation reclamation areas
                                        Perennial Ryegrass *Lolium perenne*
3 Spike congested, the broad faces of the spikelets facing the rachis; glumes
   2, subequal; lemma with a short awn 1–3 mm long
                                        Agropyron velutinum (p. 133)

4 Spikelets 1-flowered   5
4 Spikelets with 2 or more florets   7

5 Glumes truncate, the keels fringed with stiff spreading hairs and produced
   above into a short awn; panicle dense, spike-like, cylindrical; robust perennial
   ± 30–100 (–150) cm high; occasional on Soil Conservation reclamation
   areas   Timothy Grass *Phleum pratense*
5 Not as above   6

6 Lemma hyaline or membranous, thinner than the glumes   *Agrostis* (p. 122)
6 Lemma firm and chartaceous, thicker than the glumes   *Deyeuxia* (p. 124)

7 Spikelets 2-flowered, the rachilla produced into a hairy bristle   8
7 Not as above   9

8 Awn attached in the lower half of the lemma
                                        Deschampsia caespitosa (p. 128)
8 Awn attached in the upper half of the lemma   *Trisetum spicatum* (p. 127)

9 Spikelets 3-flowered, a terminal hermaphrodite floret subtended by 2 male
   florets; aromatic grass with pale brown shining spikelets
                                        Hierochloe submutica (p. 121)
9 Not as above   10

10 Lemma awnless or emarginate with a minute central point which is not
   visible to the naked eye   11
10 Lemma awned, the awn at least 1 mm long and visible to the naked eye 12

11 Lemma rounded on the back; ligule ciliate; dwarf grasses less than 10 cm
   high   *Erythranthera* (p. 128)
11 Lemma keeled; ligule membranous; otherwise not as above   *Poa* (p. 133)

12 Spikelets in dense one-sided clusters on the panicle branches; lemma keeled
   distally, the keel usually fringed with hairs; awn 1–1.5 mm long; tufted perennial ± 30–100 cm high; locally common on Soil Conservation reclamation
   areas   Cocksfoot *Dactylis glomerata*
12 Not as above   13

13 Lemma awned from the entire tip; slender perennial ± 20–40 cm high; common on Soil Conservation reclamation areas   Red Fescue *Festuca rubra*

13 Lemma notched or deeply 2-lobed at the apex, the awn attached between the lobes or below the apex of the lemma   14

---

14 Ligule a ring of hairs; perennials   15

14 Ligule membranous; annuals   16

---

15 Coarse grass forming large tussocks ± 60–90 cm high, the base of the culms surrounded by the persistent flattened brown remains of old sheaths; body of lemma with conspicuous subequal hairs evenly scattered over the back; hilum linear, about half as long as the grain   *Chionochloa* (p. 130)

15 Small tufted grasses less than 40 cm high; body of the lemma with tufts of hairs in 2 rows on the back, or the back ± glabrous; hilum elliptic, up to a quarter as long as the grain   *Danthonia* (p. 131)

---

16 Spikelets ± 5–7 cm long (incl. awns); awns ± 3–4 cm long; slender annual ± 20– 40 cm high, locally common on Soil Conservation reclamation areas, probably not persisting   Great Brome *Bromus diandrus*

16 Spikelets ± 2–2.5 cm long (incl. awns); awns ± 5–8 mm long; fairly common annual ± 15–30 cm high on Soil Conservation reclamation areas   Barley Brome *Bromus hordeaceus*

Hierochloe R. Br.

H. submutica F. Muell., Trans. Proc. Vict. Inst. 1854–55: 48 (1855)

Loosely tufted aromatic perennial 30–60 cm high; *leaves* flat or the margins slightly inrolled, ±5–25 cm × 3–8(–10) mm, acute, ribbed and scabrid on the upper surface, scabrid on the margins; *sheaths* striate, sometimes purplish, retrorsely scabrid or almost smooth; *ligule* membranous ± 5 mm long, erose or laciniate at the apex, puberulent on the back; *panicle* sparse, loosely spreading, 8–15(–18) cm long, secund or almost so, the spikelets close along the branches; *spikelets* 3-flowered, pale brown and shining, broadly elliptic, 5–6 mm long; *glumes* hyaline, acute, shining, the upper broad and subequal to or shortly exceeding the male florets, the lower narrower and slightly shorter; *rachilla* very short, sinuous; *florets* golden-brown, the 2 lower male, the upper hermaphrodite; *male florets* subequal, the lemma ± 4.5–5 mm long, hispidulous, the margins ciliate, awnless or with a minute awn ± 0.5 mm long just below the apex; *anthers* 2 mm long; *bisexual floret* slightly smaller than the male florets, the lemma unawned and glabrous except scaberulous and ± hairy in the upper third.

[71]

Alpine and subalpine tracts of the Australian Alps.

This rather rare species is found above treeline in moist sites in tall alpine herbfields and near the margins of bogs. It is sweetly coumarin-scented, and in parts of the northern hemisphere the leaves of related grasses were spread near church

Distribution

Notes and Habitat

doors because of their incense-like aroma; hence the common name, holy grasses. We suggest that an appropriate common name for this species might be Alpine Holy Grass.

References: Vickery (1975); Zotov (1973).

Agrostis L.

1 Rachilla produced into a short plumose bristle; lemma ± hairy on the
   back   *A. meionectes* (p. 122)
1 Rachilla not produced, lemma glabrous on the back   2

2 Awn exserted, conspicuous; panicle broadly pyramidal in outline with long
   capillary branches   *A. venusta* (p. 123)
2 Awn absent or rudimentary, or if present then the panicle narrow and
   contracted   3

3 Palea absent or minute (less than a quarter the length of the lemma); awn
   present or absent; tufted annuals   4
3 Palea half to two-thirds the length of the lemma; awn absent or rudimentary;
   coarse tufted rhizomatous or stoloniferous perennials   5

4 Panicle dense, contracted, with erect branches; awn present or absent
                                                         A. muellerana (p. 123)
4 Panicle spreading, pyramidal; awn absent   *A. parviflora* (p. 124)

5 Ligules of vegetative shoots shorter than wide; panicle open at maturity;
   tufted perennials usually spreading by short rhizomes or sometimes by stolons;
   common in disturbed sites and Soil Conservation reclamation areas
                                                   Brown-top Bent**A. capillaris*
5 Ligules of vegetative shoots longer than wide; panicle contracted after flowering, at least in the upper part; tufted perennial spreading by leafy stolons;
   occasional in disturbed areas   Creeping Bent **A. stolonifera*

A. meionectes J. Vickery, Contrib. N.S.W. Natl Herb. 4: 12 (1966)

[72]   Small densely tufted glabrous annual (5–) 10–30 cm high; *culms* slender, terete,
         ± retrorsely scabrid or scaberulous; *leaves* smooth or scaberulous, 1–7 cm long,
         flat, folded or almost setaceous, tapering to an acuminate tip; *ligule* membranous,
         up to 5 mm long, glabrous or puberulent on the back, acute or acuminate, becom-
         ing torn and jagged at the apex; *panicle* open and pyramidal at maturity, about
         3–8(–10) × 2–8(–10) cm, with fine, divaricate ± scaberulous branches; *spikelets*
         widely gaping at maturity; *glumes* subequal or the lower slightly longer, 3–3.5

mm long, purplish with silvery hyaline margins, becoming stramineous with age, exceeding the lemma, acute to acuminate, smooth or scaberulous on the sides, scabrid on the keels; *lemma* hyaline 2–2.5 mm long, conspicuously hairy or sometimes with only a few scattered hairs, truncate, erose and minutely toothed at the apex; *awn* straight or slightly geniculate, ± 1–3.5 mm long, attached from about half to two-thirds up the lemma; *palea* hyaline, about four-fifths as long as the lemma; *rachilla* produced into a plumose bristle 0.5–1(–2) mm long (excl. hairs); *anthers* 0.5–1 mm long.

Alpine and subalpine tracts of the Kosciusko area, N.S.W., and the Bogong High Plains and Mt Buffalo in Victoria.   *Distribution*

Rather close to *Agrostis aemula* R. Br., which occurs at lower altitudes, but the latter is larger in all its parts with a longer, more conspicuous, strongly geniculate awn; common in and on the margins of depressions in sod tussock grassland where it may form almost pure stands, and in short alpine herbfield, fen and bog communities.   *Notes and Habitat*

A. venusta Trin., Mem. Acad. Petersb. ser. 6: 340 (1841)

Delicate densely tufted glabrous annual 10–25(–40) cm high; *culms* smooth or finely scaberulous; *leaves* crowded at the base, 2–8 cm long, inrolled and almost filiform, scaberulous; *sheaths* pallid, appressed; *ligule* membranous, 2–3 mm long, obtuse, laciniate; *panicle* broadly pyramidal in outline, occupying half or more of the culm, divaricately spreading with long, verticillate, antrorsely scabrid capillary branches which are bare at their bases for 2–6 cm; *spikelets* usually purplish, 2–4 mm long; *glumes* acute, very scabrid on the keels, the lower longer than the upper; *lemma* membranous, 1.5–2.3 mm long, glabrous or the callus with a few short hairs, truncate or obscurely toothed at the apex, awned from below the middle; *awn* fine, exserted, ± 2–4 mm long, straight or slightly geniculate; *palea* minute; *anthers* 0.3–0.6 mm long.

Tablelands and ranges of eastern N.S.W., Victoria and Tasmania, and Western Australia.   *Distribution*

Not common above treeline, in tall alpine herbfields and sod tussock grassland; more common at lower elevations.   *Notes and Habitat*

Reference: Vickery (1941).

A. muellerana J. Vickery, Contrib. N.S.W. Natl Herb. 1: 103 (1941)

Tufted glabrous annual 1–20(–30) cm high; *leaves* 3–10 cm long, ± flat or folded, often setaceous and sharply keeled at the apex, scaberulous on the margins; *ligule* membranous, 2–4 mm long; *panicle* contracted, 1–6 cm long with short erect scabrid branches, exserted or the base enclosed in the expanded upper sheath; *spikelets* 2–3.5 mm long, purplish drying straw-coloured; *glumes* acute, keeled distally, scabrid on the keels, the lower longer than the upper; *lemma* ± 1.3–2   [73–4] mm long, much shorter than the glumes, hyaline, glabrous or with a few hairs on the callus, truncate, sometimes minutely toothed, unawned, or with a minute delicate awn from near the apex, or with a slender straight or geniculate exserted awn 1.5–2.6 mm long attached about the middle; *palea* minute, rarely up to one-third as long as the lemma; *anthers* 0.5–0.9 mm long.

Distribution Alpine and subalpine tracts of the Australian Alps and mountains of Tasmania.

Notes and Habitat A common grass in tall alpine herbfields, sod tussock grassland, fen and bogs. A dwarf ecotype, usually only a few cm high, is found in *Epacris–Chionohebe* feldmark.

A. parviflora R. Br., Prodr. Flor. Nov. Holl. 170 (1810)

[75] Slender glabrous erect or ascending loosely tufted annual about 10–25(–35) cm high; *culms* smooth, finely longitudinally striate; *leaves* almost flat becoming inrolled–filiform when dry, scaberulous on the margins; *ligule* membranous, obtuse, 1–2 mm long; *panicle* well exserted, ± spreading and pyramidal in outline, 3–9 cm long, the branches sparsely scaberulous or almost smooth; *spikelets* ± 1.5–2 mm long, green or purplish; *glumes* acute, shortly exceeding the lemma, subequal or the lower slightly longer, slightly keeled distally, scaberulous on the keels; *lemma* 1.5–2 mm long, membranous, glabrous, truncate, unawned; *palea* minute; *anthers* 0.3–0.7 mm long.

Distribution Tablelands and ranges of south-eastern N.S.W. (as far north as the A.C.T.) and eastern Victoria, and mountains of Tasmania.

Notes and Habitat Fairly common in depressions in sod tussock grassland, and in fens, bogs, and wet areas around pools and streams in tall alpine herbfield. Robust specimens of this species may resemble small plants of *Agrostis hiemalis* (Walt.) B.S.P., which is common in the subalpine tract and at lower altitudes. The latter, however, is a generally coarser perennial with larger panicles and conspicuous anthers 1–1.5 mm long.

Reference: Vickery (1941).

Deyeuxia Clar. ex Beauv.

KEY TO THE SPECIES

---

1 Callus hairs up to two-thirds as long or subequal in length to the lemma
D. affinis (p. 125)
1 Callus hairs less than half the length of the lemma   2

---

2 Awn more than 5 mm long, attached in the lower third of the lemma
D. monticola (p. 126)
2 Awn less than 5 mm long, attached at or above the middle of the lemma   3

---

3 Awn 0.5–2 mm long, ±straight and untwisted, attached towards the apex of the lemma; spikelets ± 4–4.5 mm long; leaves broad, ± flat or the margins slightly inrolled *D. crassiuscula* (p. 126)

3 Awn ± 4 mm long (when straightened), geniculate at maturity and loosely twisted at the base, attached slightly above the middle of the lemma; spikelets (4.5–)5–6 mm long; leaves narrow, folded, keeled distally *D. carinata* (p.126)

D. affinis M. Gray, Contrib. Herb. Aust. No. 26: 9 (1976)

Small loosely tufted perennial about (5–)10–20 cm high; *culms* slender, erect [76] or ascending and often geniculate at the lower nodes, smooth except sometimes antrorsely scaberulous below the inflorescence; *leaves* ± 1–6 cm long, moderately stiff, folded or the margins slightly inrolled, sharply callus-pointed, smooth and glabrous on the back, hispidulous on the veins of the upper surface, the margins antrorsely scaberulous; *sheaths* striate, glabrous; *ligule* membranous, 1–3 mm long, subtruncate, erose and ciliolate at the apex; *panicle* narrow and contracted, linear to narrow-elliptical in outline, ± 2–5 × 0.4–1 cm, the branches and pedicels antrorsely scabrid; *spikelets* 2.8–3.5 mm long; *glumes* subequal or the upper slightly longer, 2.5–3.5 mm long, variably tinged with dark reddish purple on the back, the upper margins silvery-hyaline, antrorsely scabrid on the keel above, acute, subacute, or with a minute excurrent mucro up to 0.3 mm long; *lemma* chartaceous becoming hyaline distally, lanceolate in outline, 2.3–2.8 mm long, 5-nerved, smooth towards the base becoming ± scaberulous especially on the nerves above, subacute or irregularly toothed at the apex; *callus* ± 0.2 mm long, blunt, densely bearded with long silky hairs, the longest of which are from two-thirds to subequal in length to the lemma; *awn* usually shortly exserted, antrorsely scabrid, straight or slightly geniculate, not or scarcely twisted, ± 1–2 mm long, attached in the upper third of the lemma about 0.5–0.8 mm from the apex; *palea* hyaline, 1.7–2.2 mm long, shorter than the lemma, subacute or very finely toothed at the apex; *rachilla* produced into a bristle 0.5–1 mm long, plumose with long hairs up to 2 mm long; *anthers* ± 0.8–1 mm long; *caryopsis* narrowly elliptical, about 1.5 mm long.

Alpine and subalpine tracts of the Mt Kosciusko area, N.S.W., and the Bogong *Distribution* High Plains, Victoria.

This rather rare species is found growing near rocks in short alpine herbfield. *Notes and* It is related to *D. aucklandica* (Hook. f.) Zotov (syn. *D. setifolia* Hook. f.) of *Habitat* New Zealand and Auckland Islands.

Reference: Zotov (1965).

D. monticola (Roem. & Schult.) J. Vickery, Contrib. N.S.W. Natl Herb. 1: 56 (1940)

Erect tufted perennial 10–40(–70) cm high; *leaves* 5–20 cm long, inrolled, usually subulate, with a sharp callus tip, strongly antrorsely scabrid or almost smooth; *ligule* membranous, obtuse, ciliate at the apex, up to 2 mm long; *panicle* contracted, linear to linear-lanceolate, 5–15 cm long; *spikelets* compressed, 4.5–7 mm long; *glumes* subacuminate, green or purple-tinged with hyaline margins, strongly keeled, scabrid on the keels, subequal or usually the lower slightly shorter than the upper; *lemma* 4.5–6 mm long, densely scaberulous, 4-nerved distally, with two of the nerves produced into minute teeth at the apex, the callus hairy; *awn* conspicuously exserted, attached below the middle, geniculate, twisted below the bend, 5.5–10 mm long (when straightened); *palea* shorter than or occasionally subequal to the lemma, with 2 minute teeth at the apex; *rachilla* produced into a glabrous or hairy bristle.

Distribution	Tablelands and ranges of eastern N.S.W. and Victoria from the Barrington Tops, N.S.W., to Mt Macedon, Vic., and mountains of Tasmania.
Notes and Habitat	A rather variable species, not common above treeline, in sod tussock grassland and tall alpine herbfield.

D. crassiuscula J. Vickery, Contrib. N.S.W. Natl Herb. 1: 59 (1940)

[77] Coarse, loosely tufted, erect or ascending perennial to 45 cm high; *leaves* stiff and suberect, thick, flat or the margins slightly inrolled, 3–12 cm × 3–8 mm, narrowed to an acute point, smooth or minutely scaberulous on the back, conspicuously ribbed and scabrid to pubescent on the upper surface; *sheaths* striate, scarcely scaberulous, often deeply stained with reddish purple; *ligule* membranous, truncate, 1–3 mm long, ± densely puberulent on the back, laciniate and shortly ciliate at the apex; *panicle* contracted, dense and spike-like, shining, green- and purple-tinted, narrow lanceolate in outline or subcylindrical, about 6–10 × 0.5–1.5 cm; *spikelets* ± 4–4.5 mm long, ± turgid; *glumes* ± 4–4.5 mm long, keeled, acute or minutely mucronate, minutely scaberulous on the sides, the keel scabrid above; *lemma* firm and indurated, microscaberulous, conspicuously 5-nerved towards the base, subequal to or slightly shorter than the glumes, glabrous except for the conspicuously bearded callus, with a small, weak ± straight untwisted awn 0.5–2 mm long attached towards the apex; *palea* subequal to or slightly shorter than the lemma; *rachilla* produced into a ± glabrous bristle about 0.7 mm long; *anthers* ± 1–1.4 mm long.

Distribution	Alpine and subalpine tracts of the Australian Alps as far north as the Brindabella Range, A.C.T.
Notes and Habitat	Fairly common in tall alpine herbfields and sod tussock grassland, often near large rock outcrops.

D. carinata J. Vickery, Contrib. N.S.W. Natl Herb. 1: 58 (1940)

[78] Superficially similar to the preceding (*D. crassiuscula*), differing mainly as follows: *leaves* narrower, less rigid, flat or folded, keeled distally, with a sharp or subobtuse

callus point; *ligule* sometimes up to 6 mm long; *spikelets* compressed, usually larger, (4.5–)5–6 mm long; *lemmas* more compressed, smooth and scarcely nerved at the base, coarsely scaberulous above, with a more conspicuously exserted geniculate awn ± 4 mm long, loosely twisted at the base and attached slightly above the middle of the lemma; the spikelets are apparently cleistogamous with small anthers 0.4–0.7 mm long.

Distribution

As for *D. crassiuscula*, also Mt Field, Tasmania.

Notes and Habitat

Common and widespread in wet areas in sod tussock grassland, fen, bogs and heaths, and near streams and pools.

Trisetum Pers.

T. spicatum (L.) Richt., Pl. Eur. 1: 59 (1890)   'Bristle-grass'

[79–81]

Very variable tufted perennial 20–40 cm high; *culms* densely velvety-pubescent, often retrorsely so below and antrorsely above, or almost glabrous; *leaves* flat or folded, ± 5–15 × 0.2–1 cm, the upper surface strongly ribbed and ± scabrid-pubescent, the underside puberulent to almost glabrous or sometimes with sparse long hairs, the margins scabrid, sometimes with sparse long hairs; *upper sheaths* usually densely pubescent, the lower puberulent or subglabrous; *ligule* membranous, 1–2 mm long, torn and ciliate at the apex; *panicle* ± 4–10(–15) × 0.7–2 cm, contracted and spike-like at maturity (see Notes below), ± interrupted towards the base; *spikelets* shining, compressed, 2-flowered, ± 5–7 mm long; *glumes* 4–6 mm long, membranous, slightly shorter than the florets, green- and purple-tinted with hyaline margins, the lower shorter and narrower than the upper, the keel scabrid; *lemma* 5–6.5 mm long, scaberulous, the margins hyaline, bifid or minutely 2-toothed at the apex, the callus shortly bearded; *awn* 3–6 mm long, well exserted, recurved or sometimes weakly geniculate, attached about one-third below the apex of the lemma; *palea* hyaline, bifid at the apex, usually slightly shorter than the lemma; *rachilla* hairy, produced into a bristle 1–2 mm long; *anthers* 1.5–2.5 mm long.

Distribution

South-east Queensland, alpine and subalpine tracts of the Australian Alps, and mountains of Tasmania and New Zealand; a widespread species in mountainous areas of the world.

Notes and Habitat

Australian and New Zealand forms of this widespread species have been described as subsp. *australiense* Hultén. The appearance of this grass changes markedly at anthesis when the panicle opens out and becomes ovate or elliptical in outline. Typical habitats are damp areas in tall alpine herbfields and sod tussock grassland. Reference: Hultén (1959).

Deschampsia Beauv.

D. caespitosa (L.) Beauv., Ess. Agrost. 91, t. 18, fig. 3 (1812)

'Tufted hair-grass'

[82] Tufted or tussock-forming glabrous perennial (15–)30–90 cm high (sometimes taller at lower altitudes); *leaves* shorter than the culms, flat, folded, or inrolled when dry, stiff and sharp-pointed, smooth on the underside, coarsely ribbed on the upper surface, the ribs and margins antrorsely scabrid; *sheaths* pale green, smooth and glabrous; *ligule* membranous, glabrous, acute, 3–8 mm or more long; *panicle* 10–20(–30) cm or more long with slender whorled branches, contracted and drooping at anthesis; finally ± erect, loose and open; *spikelets* 2-flowered, ± 6 mm long; *glumes* membranous, subequal, 3–6 mm long, acute, not or scarcely overtopping the florets, green and purple-tinted proximally, silvery and shining distally; *rachis* conspicuously hairy and produced into a short plumose bristle; *lemma* membranous, 3–4 mm long, truncate and toothed at the apex, glabrous except for the conspicuously bearded callus, with a slender straight awn up to 4 mm long from near the base; *palea* hyaline, subequal to or slightly shorter than the lemma; *anthers* 1.5–2 mm long.

Distribution   Alpine and subalpine tracts of the Australian Alps and associated ranges, the Mt Lofty Range, South Australia, and Tasmania; widespread in temperate and arctic regions of the world and in mountainous areas of the tropics.

Notes and   This cosmopolitan grass has a very scattered distribution above treeline, and
Habitat   is easy to recognize when in flower on account of the shining purple and silvery panicles. It occurs in wet areas, sometimes growing in several centimetres of water as in seepages, pools and along stream margins; also in sod tussock grassland.

Erythranthera Zotov

KEY TO THE   1   Spikelets 2.5–3.8 mm long; glumes ovate, subobtuse; lemma glabrous
SPECIES   *E. australis* (p. 128)
            1   Spikelets 4.5–5.2 mm long; glumes lanceolate, acute; lemma ± hairy
                                                 E. pumila (p. 129)

E. australis (Petrie) Zotov, N.Z. J. Bot. 1: 125 (1963)

[83–4] Dwarf perennial 1–10 cm high, from slender, branched, erect or ascending rhizomes, sometimes forming a dense short sward; *culms* glabrous, slender, erect or geniculate at the base, smooth, terete, subequal to or overtopping the leaves; *leaves* short, stiff, 1–3 cm long, folded or inrolled – setaceous, smooth and glabrous;

sheaths membranous, glabrous, striate, sometimes bearded at the orifice, the lower sheaths pallid, the upper sheaths green- or purple-tinged and occasionally obscurely red-banded; *ligule* ciliate, ± 0.3 mm long; *inflorescence* short, racemose or subracemose, 0.5–1.3 × ± 0.3 mm, the spikelets 2–5(–7), on short erect pedicels ± appressed to the rachis; *spikelets* 2–3-flowered, rather plump, asymmetrical, ± 2.5–3.8 mm long, the florets shortly exceeding the glumes; *glumes* persistent, ovate, subobtuse, rounded on the back, slightly keeled above, green- and purple-tinged, the upper subequal to or slightly shorter than the lower and ± enfolded by it at least at the base; *lemma* (incl. the short blunt callus) glabrous, 1.6–2.2 mm long, suborbicular, obtuse or emarginate with a minute central bristle up to 0.2 mm long in the notch, rounded on the back, sometimes obscurely keeled above; *palea* narrowly obovate, emarginate, glabrous except ciliate at the apex and on the keels above, subequal to or slightly shorter than the lemma; *anthers* ± 0.5 mm long.

Alpine tract (and from Betts' Camp in the subalpine tract) of Kosciusko area, N.S.W., the Bogong High Plains, Victoria, and Mt Field, Tasmania; also main divide of New Zealand (North and South Islands). *Distribution*

This small, locally common but apparently sparsely distributed grass was overlooked at Kosciusko until 1958 when it was first collected in the area in short alpine herbfield and on wet gravelly creek banks. Earlier livestock grazing may have reduced its abundance, as it did with other palatable herbs such as *Chionochloa* and species of *Ranunculus*, *Aciphylla*, etc. In this connection Petrie (1890) in his original description of the grass from New Zealand comments: 'It is a nutritious grass, much relished by sheep and horses, and usually closely cropped.' *Notes and Habitat*

References: Blake (1972); Gray (1974); Petrie (1890).

E. pumila (Kirk) Zotov, N.Z. J. Bot. 1: 124 (1963)

Dwarf loosely tufted perennial 1–7 cm high; *culms* slender, almost filiform, glabrous, erect or ascending with persistent remains of old sheaths at the base; *leaves* short, stiff, smooth and glabrous, ±0.5–3 cm long, inrolled-setaceous, often curved or ± spirally twisted; *sheaths* striate, ± bearded at the orifice, glabrous or pilose on the back with glabrous membranous margins, the upper sheaths purple-tinged or occasionally with obscure reddish purple lateral bands; *ligule* ciliate, up to 0.3 mm long; *inflorescence* racemose or subracemose, up to 1.5 cm long; *spikelets* 2–4 or occasionally solitary, erect and ± appressed to the rachis, 3–4-flowered, 4.5–5.2 mm long, the glumes exceeding the florets; *glumes* lanceolate, acute, rounded on the back, slightly keeled distally, green- and purple-tinged with scarious margins, the upper subequal to or shortly exceeding the lower and enfolded by it at the base; *lemma* ovate, 2–2.5 mm long (incl. the blunt glabrous callus), with hairs scattered over the back, emarginate with a minute central bristle up to 0.3 mm long in the notch; *palea* glabrous except ciliate on the keels, emarginate, subequal to or shortly exceeding the lemma; *anthers* ± 0.5 mm long. [85]

Mt Northcote and Mt Lee, Kosciusko area, N.S.W., and eastern mountains of the South Island of New Zealand. *Distribution*

This grass, first collected in *Epacris–Chionohebe* feldmark on Mt Northcote in 1965†, and still only known from the area between Mt Northcote and Mt Lee, is probably one of the rarest and most restricted grasses in Australia. Even scientific botanical collection should be strictly limited in order to conserve this species.

References: Blake (1972); Gray (1974); Zotov (1963).
†Earlier find revealed: C. Scottsberg, 11.iii. 1949, near Lake Albina.

Chionochloa Zotov

C. frigida (J. Vickery) Conert, Senckenbergiana Biol. 56: 154 (1975)   'Ribbony grass'

[86–8]   Very robust perennial about (45–)60–120 cm high, usually forming large dense tussocks up to 1 m or more in diam.; *culms* glabrous, smooth and shining, shorter than or shortly overtopping the leaves, surrounded at the base by the persistent ± curved brittle remains of the broad flattened truncated sheaths; *leaves* flat or folded and up to 10 mm wide at the base, gradually tapering to a long subulate-pointed tip which is often 30–60 cm long or more, strongly inrolled on drying, smooth and shining on the underside, the deep grooves of the upper surface microscaberulous, the margins distantly scabrid and usually with sparse long hairs towards the base; upper culm leaves finer and narrower, sometimes overtopping the panicle; *sheaths* of the basal leaves loose, straw-coloured, 10–13 mm wide, conspicuously striate, with narrow chartaceous margins; *ligule* a ring of short cilia 0.5–1.5 mm long; *panicles* ± 9–15 × 3–8 cm, only shortly exserted from the uppermost sheaths, nodding at maturity, the branches antrorsely hirsute; *spikelets* flattened, 1–1.8 × 0.5–0.7 cm, usually 5–7-flowered, the florets usually exceeding the glumes; *glumes* 8–13 mm long; *lemma* uniformly hairy all over the back, the body 4–7 mm long incl. the ± 1 mm conspicuously hairy callus, the lateral lobes subequal in length to the body, with membranous margins, and usually acuminate with short awns up to 3 mm long; *central awn* exserted, 8–15 mm long, geniculate, flattened and slightly twisted at the base; *palea* 7–9 mm long, linear to linear-lanceolate, glabrous on the back but densely pilose between the ciliate keels and the margins in the central part; *anthers* 2.5–4 mm long; $2n = 42$ (as *Danthonia frigida*)*.

Distribution   Endemic to the alpine and upper subalpine tracts of the Kosciusko area, N.S.W.

Notes and
Habitat   *C. frigida* forms distinctive communities within the more widespread *Celmisia–Poa* herbfields. The largest stands now occur on the steep, less accessible western slopes of the Main Range in areas of granite. Formerly it was also common on the eastern slopes of Kosciusko (Helms 1893; Maiden 1898, 1899), but with grazing and burning it almost became extinct there by the 1930s (Costin 1958). Since protection of the Kosciusko summit area from grazing in 1944, *C. frigida* has been recovering. Work on New Zealand species of *Chionochloa* (known as snow grasses in that country) indicate that they are both slow-growing and long-lived; individual tussocks may reach well over 100 years of age.

References: Blake (1972); Brock and Brown (1961)*; Costin (1958); Helms (1893); Maiden (1898, 1899); Vickery (1956); Zotov (1963).

Danthonia DC.

These three species would be included in the genus *Notodanthonia* Zotov in the sense of Zotov (1963) and Blake (1972); however, as this genus has not been taken up in any regional Australian flora, I have refrained from making the necessary combinations here to avoid nomenclatural confusion.

KEY TO THE SPECIES

1 Lemma with two conspicuous rows of hairs across the back; panicle ovate to ovate-lanceolate in outline, ± 2–3 cm wide, the awns well exserted and conspicuous   *D. alpicola* (p. 131)
1 Lemma with marginal tufts of hairs, but the back glabrous or with only inconspicuous hairs or tufts developed; panicle linear or narrow-lanceolate in outline, usually less than 1.5 cm wide, the awns not or only shortly exserted   2

2 Spikelets (excl. awns) ± 1–1.3 cm long; lateral lobes longer than the body of the lemma, the central awn 6–8 mm long, shortly exserted   *D. nudiflora* (p. 132)
2 Spikelets (excl. awns) ± 5–7 mm long; lateral lobes usually shorter than the body of the lemma, the central awn 1.5–3 mm long, not or scarcely exserted   *D. nivicola* (p. 132)

D. alpicola J. Vickery, Contrib. N.S.W. Natl Herb. 1: 297 (1950)
'Crag wallaby-grass'

Loosely tufted perennial 10–30 cm high, glabrous except often with marginal tufts of hairs at the orifice of the sheaths; *culms* smooth, erect or ascending and ± geniculate at the nodes; *leaves* smooth, rigid, spreading, 5–10 cm long, closely folded or inrolled or sometimes almost flat towards the base, keeled and pungent at the apex; *sheaths* smooth, rather thick, often purple-tinted proximally, striate distally; *panicle* dense, contracted, about 3–5 × 2–3 cm, ovate or ovate-lanceolate in outline; *spikelets* 4–6-flowered, purplish becoming straw-coloured, the glumes exceeding the florets except for the long-exserted awns; *glumes* acute, subequal, 12–17 mm long; *lemma* with two conspicuous rows of hair-tufts across the back extending to the margins, the body 3–3.5 mm long incl. the blunt hairy callus, the lateral lobes membranous at the base for about 3–5 mm, narrowing above into setae about 3–5 mm long; *central awn* geniculate, twisted at the base, 10–17 mm long and much exceeding the lateral lobes; *palea* about 5.5–7 mm long, linear-lanceolate, glabrous on the back or occasionally hairy below the middle; *anthers* about 1.5 mm long.

Alpine and subalpine tracts of the Australian Alps and Mt Murray, A.C.T.

In crevices and on ledges of crags and large rock outcrops as a dominant species of *Brachycome–Danthonia* tall alpine herbfield.

Reference: Vickery (1956).

[89]

Distribution

Notes and Habitat

D. nudiflora P. F. Morris, Victorian Nat. 52: 111, t. 10 (1935)
'Alpine wallaby-grass'

[90]   Erect densely tufted perennial 10–40 cm high; *culms* stiff, glabrous except sometimes puberulent just below the inflorescence, ± densely covered with old sheaths
at the base; *leaves* smooth, glabrous, 2–10(–15) cm long, inrolled and more or
less subulate, stiff and subpungent; *panicle* contracted, about 3–7 × 0.5–1.5 cm,
linear or narrow-lanceolate in outline; *spikelets* 5–6-flowered, about 1–1.3 cm
long excl. the shortly exserted awns; *glumes* 8–13 mm long, acute, slightly keeled
distally, subequal or the lower slightly exceeding the upper; *lemma* glabrous on
the back with two tufts of hair on each margin, or occasionally with two inconspicuous dorsal tufts of hairs, the body about 2.5–3 mm long incl. the short
hairy callus, the lateral lobes about 4–5 mm long, flat with membranous margins
at the base, tapering above into 2–2.5 mm setae; *central awn* 6–8 mm long, loosely
twisted at the base, much exceeding the lateral lobes; *palea* 4–5 × 1–1.5 mm,
oblong-lanceolate, glabrous on the back, puberulent on the inner surface in the
upper part; *anthers* ± 1–1.8 mm long; $2n = 24^*$.

Distribution   Alpine and subalpine tracts of the Australian Alps and associated ranges as far
north as Mt Gingera in the A.C.T., and mountains of Tasmania.

Notes and   A dominant species of sod tussock grassland; also in wet areas of tall alpine herb-
Habitat   field and in fens.

References: Brock and Brown (1961)*; Vickery (1956).

D. nivicola J. Vickery, Contrib. N.S.W. Natl Herb. 1: 300 (1950)
'Snow wallaby-grass'

[91]   Small densely tufted perennial about 10–25(–30) cm high; *culms* very slender,
stiff, erect or oblique, smooth and glabrous, usually reddish; *leaves* very fine,
smooth, setaceous, tightly inrolled, up to 10 cm long, glabrous or sometimes
with a few hairs on the inner surface; *sheaths* smooth, glabrous or with small
marginal tufts of hairs at the orifice; *panicle* linear or narrow-lanceolate in outline,
subracemose, 1.5–4.5 × 0.4–1 cm; *spikelets* usually 4-flowered, ± 5–7 mm long,
purplish drying straw-coloured; *glumes* acute, 4–7 mm long, slightly keeled distally, subequal to or slightly exceeding the florets; *lemma* 3–4 mm long, glabrous
on the back except for scattered hairs towards the base, two conspicuous tufts
near each margin and sometimes two tufts near the sinus, the lateral lobes usually
shorter than the body of the lemma, acute, or acuminate with very short setae
up to 0.5 mm long; *central awn* straight or slightly twisted at the base, 1.5–3
mm long, shortly exceeding the lateral lobes; *palea* 2.5–3 × 1 mm; *anthers* ±
1 mm long; $2n = 24, 48^*$.

Distribution   Alpine and subalpine tracts of the Australian Alps and mountains of Tasmania.

Notes and   Often associated with *Danthonia nudiflora* in sod tussock grassland, in which it
Habitat   is locally conspicuous because of the masses of shining reddish stems; also in damp
sites in tall alpine herbfields, fen, etc.

References: Brock and Brown (1961)*; Vickery (1956).

Agropyron Gaertn.

A. velutinum Nees, Hook. Lond. J. Bot. 2: 417 (1843)
'Velvet wheat-grass'

Loosely tufted perennial about 10–30(–60) cm high; *culms* stiff, erect or ascending [92]
and ± geniculate towards the base, pubescent with retrorse-appressed hairs, or
sometimes with spreading or antrorse hairs below the inflorescence; *leaves* with
moderately stiff suberect blades ± 5–12(–18) cm long, flat or inrolled, strongly
ribbed and ± scabrid-pubescent on the upper surface, glabrous, pubescent or
scaberulous on the underside, the margins scabrid; *sheaths* pubescent or puberulent, often with small pointed auricles at the apex; *ligule* a short truncate ±
ciliate membrane ± 0.2 mm long; *spike* rigid, congested, or ± interrupted below,
about 1.5–6 × 0.7–1.5 cm, the rachis notched and pubescent; *spikelets* stiff, mostly
6–8-flowered, green with purplish tips, sessile in 2 rows on opposite sides of
the rachis and ± appressed to it; *glumes* persistent, indurated, conspicuously
nerved, 5–7.5 mm long incl. the pungent awn-like tips, obscurely puberulent
on the inside; *lemmas* indurated, lanceolate, ± 7–10 mm long, smooth or scaberulous, rounded on the back at the base, keeled with a prominent scabrid mid-nerve
above which is produced into a 1–3 mm rigid awn; *palea* ± 6–7 × 1.5 mm,
the keels conspicuously fringed with short stiff bristles above; *anthers* 2–3 mm
long; *grain* ± 4 × 1–1.5 mm, pubescent at the apex.

Alpine and subalpine tracts of the Australian Alps and associated ranges as far   *Distribution*
north as Mt Gingera, A.C.T., and mountains of Tasmania.

Fairly common in tall alpine herbfields and sod tussock grassland.   *Notes etc.*

Poa L.

Measurements of lemma length refer to the lowermost lemma of the spikelet;   KEY TO THE
the lemmas of several spikelets from each specimen should be examined. The   SPECIES
key to *P. costiniana*, *P. fawcettiae* and *P. hiemata* can only be regarded as provisional.
The genus *Poa* is very difficult taxonomically (Vickery 1970), and combinations
of characters rather than single characters must often be used to separate the
species. Intermediate populations, e.g. between *P. costiniana* and *P. fawcettiae*,
appear to be common above the treeline – see notes under the individual species.

---

1   Small weak annual usually less than 15 cm high, with soft, flat or folded
     leaves less than 10 cm long; disturbed areas only
                                   Annual Meadow-grass*P. annua
1   Perennials, not as above   2

---

2   Panicle contracted, linear with short erect branches and plump spikelets;
     lemma broad, cymbiform, glabrous on the back; palea ovate or elliptical,
     densely and conspicuously long-ciliate on the keels above
                                                      P. saxicola (p. 134)

2   Panicle loose, usually spreading and pyramidal at maturity, the spikelets ± flattened; lemma narrow, hairy or pubescent on the back or at least on the keel, or if sometimes glabrous (in *P. costiniana*) then the palea narrow and scabrid or sparsely ciliate on the keels above   3

---

3   Plants with slender creeping rhizomes; leaves soft, flat or folded; disturbed areas only   Kentucky Blue-grass*P. pratensis*
3   Densely tufted plants often forming extensive swards; leaves convolute or closely folded and compressed-terete   4

---

4   Lemmas hairy chiefly on the keel and marginal nerves and usually glabrous between or the lemma devoid of hairs, 3.5–5.5 mm long but mostly ± 4–4.5 mm long; leaves usually coarse, green, rigid and pungent, the lower sheaths mostly pallid   *P. costiniana* (p. 135)
4   Lemmas ± pubescent all over the lower part of the back, 2–3.5(–4) mm long   5

---

5   Lemmas (3–)3.5(–4) mm long; leaves moderately coarse, stiff and usually bluish, the sheaths often deeply stained with reddish purple
                                              P. fawcettiae (p. 135)
5   Lemmas (2–)3(–3.5) mm long; leaves fine and soft, usually bright green, at least the outer sheaths usually pallid   *P. hiemata* (p. 136)

P. saxicola R. Br., Prodr. Flor. Nov. Holl. 180 (1810)   'Rock poa'

[93]   Glabrous densely tufted perennial about 20–60 cm high from a vertical or oblique rootstock; *culms* slender, usually smooth, terete or slightly compressed; *sheaths* compressed, smooth or occasionally slightly scabrous, pale becoming brownish with age; *leaves* erect, shining, 10–25(–30) cm long, ± expanded and up to 4 mm wide when fresh, closely folded when dry, usually smooth to the touch or the margins antrorsely scaberulous; *ligule* membranous, 1–4.5 mm long, truncate and jagged-ciliate at the apex; *panicle* narrow and contracted, ± linear, 5–10(–15) cm long, overtopping the leaves and often nodding at maturity; *spikelets* plump, broadly elliptical in outline, ± 6–7 × 2–3 mm, few flowered, the florets closely imbricate; *glumes* broadly ovate, cymbiform, acute or subobtuse, more than half the length of the spikelet; *lemma* broadly ovate, cymbiform, 4–5.5 mm long, firm and subchartaceous, glabrous except sometimes with a few short hairs on the callus and ciliate on the membranous margins proximally; *palea* broad, firm, ovate or elliptical, usually shorter than the lemma, glabrous on the back or shortly pubescent upwards, the keels densely and conspicuously ciliate except at the base, the cilia in more than 1 row; *anthers* 0.75–1.5 mm long.

Distribution   Sparsely distributed in the alpine and subalpine tracts of the Australian Alps and associated ranges as far north as the Brindabella Range, A.C.T., and high mountains of Tasmania.

Notes and Habitat   Mainly in tall alpine herbfield, often in the shelter of rocky outcrops. It can easily be distinguished from other species of *Poa* in the area by the plump spikelets and the narrow contracted panicles which usually droop at maturity.

Reference: Vickery (1970).

P. costiniana J. Vickery, Contrib. N.S.W. Natl Herb. 4: 214 (1970)

Densely tufted perennial ± 15–40(–80) cm high; *leaves* rather rigid and pungent, [94]
convolute and terete or closely folded and compressed, finely scabrid-puberulent
on the inner surface, usually bright green and shining, smooth to the touch
or sometimes scabrous-pubescent proximally above the sheath; *sheaths* smooth
or sparsely retrorsely scaberulous, the lower usually pallid or very rarely purplish,
the upper sometimes pale reddish purple; *ligule* ± 0.5–3 mm long, membranous,
truncate-obtuse, ciliolate at the apex and puberulous on the back; *panicle* loose,
usually pyramidal or elliptical in outline, about 5–10(–20) cm long; *spikelets*
2–5-flowered, ovate in outline; *lemma* (3.5–)4–4.5(–5.5) mm long, hairy proximally on the keel and marginal nerves and glabrous or almost so between, or
occasionally wholly glabrous, the web hairs on the callus variable, from well
developed to absent; *palea* narrow lanceolate or narrowly elliptical, glabrous or
sometimes ± pubescent on the back, the keels scabrid or sparsely ciliate except
at the base; *anthers* ± 2–2.6 mm long.

Alpine and subalpine tracts of the Australian Alps and associated ranges as far   *Distribution*
north as the Brindabella Range, A.C.T., with an isolated form on the New
England Tablelands, N.S.W.

P. costiniana is the main dominant in sod tussock grassland. It also occurs in wet   *Notes and*
areas in tall alpine herbfields, in fens and bogs and along streams. It can be dis-   *Habitat*
tinguished from *P. hiemata* in the field by its larger, heavier-looking panicles
and spikelets and coarser, more rigid, ± pungent leaves (see note under *P.
hiemata*); however, populations intermediate between *P. costiniana* and *P. fawcettiae* appear to be common above treeline. *P. costiniana* is very closely related
to the Tasmanian species *P. gunnii* J. Vickery. We suggest that the common
name Prickly Snow Grass would be appropriate for this species.

P. fawcettiae J. Vickery, Contrib. N.S.W. Natl Herb. 4: 232 (1970)
'Smooth-blue snow grass'

Glabrous, densely tufted perennial ± 20–60 cm high; *leaves* usually bluish green   [95]
and glaucous (at least when dry), moderately stiff and pungent-pointed, tightly
folded and subterete or ± convolute, finely scabrid-puberulent on the inner
surface, smooth to the touch, sometimes minutely scabrid proximally; *sheaths*
typically purplish fading to pale brown, rather closely folded; *ligule* membranous,
0.5–1.5 mm long, obtuse to truncate, puberulous on the back; *panicles* ±
3–10(–18) cm long, loose and pyramidal or elliptical in outline at maturity;
spikelets compressed, 3–5-flowered; *lemma* oblong, (3–)3.5(–4) mm long, pubescent on the back in the lower part, usually with some longer hairs on the
keel and marginal nerves, the web hairs on the callus present or absent; *palea*
narrow, pubescent on the back, scabrid and/or minutely ciliate on the keels;
anthers ± 2 mm long.

Alpine and subalpine tracts of the Australian Alps and associated ranges from   *Distribution*
the Brindabella Range in the A.C.T. to the Baw Baws in Victoria, with an
outlier in the Grampians.

Notes and Habitat This species is more common in the subalpine tract where its bluish leaves, purplish outer sheaths and smaller spikelets distinguish it from *P. costiniana*. However, above treeline, *P. fawcettiae* tends to have larger spikelets, and other characters also tend to intergrade with those of *P. costiniana*. *P. fawcettiae* is an important component of tall alpine herbfields, but also occurs in other communities including drier areas in sod tussock grassland.

P. hiemata J. Vickery, Contrib. N.S.W. Natl Herb. 4: 230 (1970)
'Soft snow grass'

[96] Glabrous densely tufted perennial about 15–60 cm high; *leaves* smooth to the touch or almost so, very fine and soft, green and shining when fresh, closely folded or convolute and ± compressed-terete, scabrid-puberulent on the inner surface; *sheaths* thin and ± compressed, smooth or almost so, pallid or the upper pale reddish purple; *ligule* ± 0.5 mm long, membranous, truncate, ciliolate at the apex, scaberulous on the back; *panicle* 5–12 cm long, pyramidal in outline at maturity; *spikelets* compressed, ovate, purplish or greenish, 3–5-flowered; *lemma* (2–)3(–3.5) mm long, acute to obtuse, pubescent all over the back in the lower part, sometimes with longer hairs on the keel and marginal nerves, the web hairs on the callus sparse or absent; *palea* narrow, the keels ciliolate below and scaberulous above, pubescent on the back in the lower half; *anthers* 1.5–2 mm long.

Distribution Alpine and subalpine tracts of the Australian Alps and associated ranges as far north as the Brindabella Range, A.C.T.

Notes and Habitat The common name of the species, Soft Snow Grass, distinguishes it from its congeners, *P. costiniana* and *P. fawcettiae,* both of which tend to have coarser and more rigid leaves and larger spikelets. The contrast is particularly apparent in the case of *P. costiniana,* and one cannot do better than quote Vickery (1970, p. 217): 'On moist alpine meadow slopes *P. costiniana* and *P. hiemata* often form a dense mixed turf in which it may at times be difficult to discriminate between the two species in the field; however, the leaves of *P. costiniana* are generally coarser and more rigid, while the panicles and spikelets are generally coarser and heavier-looking than those of *P. hiemata*. Although curious as a botanical test, it is perhaps none the less worth recording that upon a pure stand of *P. hiemata* the collector will find a soft and enjoyable resting place, while on a pure stand of *P. costiniana* it will prove more prickly and uncomfortable.'

Cyperaceae

KEY	1	Flowers bisexual, nut not enclosed in a utricle	2
	1	Flowers unisexual, nut enclosed in a utricle	5

2	Hypogynous bristles or scales absent	*Scirpus* (p. 137)
2	Hypogynous bristles or scales present	3

3	Hypogynous scales remaining attached to the top of the peduncle after the nut has fallen	*Oreobolus* (p. 142)
3	Hypogynous bristles attached to the base of the nut and falling with it	4

4	Hypogynous bristles conspicuous, plumose	*Carpha* (p. 140)
4	Hypogynous bristles small and inconspicuous, scaberulous	*Schoenus calyptratus* (p. 141)

5	Rachilla produced into a prominent exserted hook	*Uncinia* (p. 143)
5	Rachilla not produced	*Carex* (p. 145)

Scirpus L.

KEY TO THE SPECIES

1 Style branches 2 — *S. crassiusculus* (p. 138)
1 Style branches 3 — 2

2 Stamens mostly 3 per floret, rarely 1–2 in the upper florets; spikelet 1, suberect, partly enfolded by the broadly dilated base of the erect floral bract which is shorter than the culm — *S. aucklandicus* (p. 138)
2 Stamens 1 per floret, or if 2–3 in the lower florets then the spikelets 1–3, the base of the floral bract not as widely dilated as the above, and/or the floral bract longer than the culm — 3

3 Rhizomes branched and long-creeping; glumes usually blunt or emarginate (occasionally acute), subequal to or only shortly exceeding the nut; stamen 1 (rarely 2–3 in lowermost florets); nut 0.7–1(–1.1) mm long — *S. subtilissimus* (p. 139)
3 ± Tufted plants, the rhizomes, if present, ascending or only shortly creeping; glumes acute or with the keel shortly excurrent, distinctly and regularly exceeding the nut; stamens usually varying within the spikelet from 3–2 in the lower florets to 1 above; nut (0.9–)1–1.4 mm long — 4

4 Nut obovate-elliptic to broadly elliptic in outline, trigonous with the dorsal angle distinctly ribbed; floral bracts shorter than the culms, the spikelets usually strongly tinged with red-purple — *S. habrus* (p. 139)
4 Nut obovate to obovate-elliptic in outline, the dorsal angle not or scarcely ribbed; floral bracts longer than the very short culms, the spikelets ± hidden among the foliage, the glumes greenish or sometimes narrowly tinged with red-purple towards the margins — *S. montivagus* (p. 139)

S. crassiusculus (Hook. f.) Benth., Flor. Aust. 7: 326 (1878)

[97] Perennial with erect or ascending culms 2–8(–15) cm long, forming tufts or dense short swards on mud near the water's edge, sometimes sending floating stems into the water; *rhizomes* branched, ascending, or long-creeping and rooting at the nodes; *leaves* narrow-linear, blunt, 2–6 cm × 0.5–0.8 mm, channelled proximally, ± flattened distally; *sheaths* membranous, loose, pale green or pallid, drying pale brown; *spikelet* plump, solitary and terminal, ovoid, acute, 5–8 × 2.5–5 mm, pale green ± flecked with red, ebracteate or sometimes with a short bract rarely exceeding the spikelet; *glumes* ovate, obtuse, finely nerved, ± twice as long as the nut, green at the centre, the margins membranous, usually stained with reddish purple; *stamens* 3; *anthers* ± 1 mm long; *style branches* 2; *nut* compressed, 1.4–2 × 0.8–1.4 mm, grey or pale brown, obovate, oval or orbicular, shortly apiculate, smooth, with slightly thickened margins.

Distribution Alpine and subalpine tracts of the Australian Alps and associated ranges as far north as the A.C.T., and mountains of Tasmania; also New Zealand (North Island), high mountains of New Guinea and from one locality in Japan.

Notes and Habitat Not common above treeline, but occasionally found in wet areas near lakes and streams and in depressions in short and tall alpine herbfields.

Reference: Kern (1974).

S. aucklandicus (Hook. f.) Boeck., Linnaea 36: 491 (1870)

[98] Perennial 1–7 cm high, from slender branched ascending or long-creeping rhizomes, tufted through *Sphagnum* or forming dense short turf in wet mud; *leaves* well developed, stiff, channelled, acute or subobtuse, narrow-linear or almost setaceous, up to about 0.5 mm wide, usually overtopping the culms; *sheaths* membranous, finely nerved, truncate at the orifice, hyaline becoming brown with age, minutely red-flecked or sometimes deep reddish purple; *spikelet* solitary, ovoid, suberect, partly enfolded by the dilated base of the erect floral bract; *glumes* ovate, deciduous, 1.2–2 mm long, acute, pale green to pale brown, minutely red-flecked or wholly deep reddish purple, contracted above, the keel thickened and ± excurrent at the apex; *stamens* 3; *anthers* 0.5–0.8 mm long, distinctly apiculate; *style branches* 3; *nut* 1.1–1.3 × 0.6–0.9 mm, obovate-elliptic, apiculate, trigonous with slightly ribbed angles, smooth and shining, stramineous becoming greyish at maturity.

Distribution Alpine and subalpine tracts of the Australian Alps and associated ranges as far north as the A.C.T., and mountains of Tasmania; also New Zealand (North and South Islands), Stewart, Antipodes, Auckland, Campbell, Macquarie and Amsterdam Islands, and New Guinea (Mt Hagen, 3350 m altitude, and Mt Wilhelm).

Notes etc. Short alpine herbfield, wet areas in bogs and fen and along lakes and streams.

References: Blake (1947, 1969); Kern (1974).

S. *subtilissimus* (Boeck.) S. T. Blake, Contrib. Queensl. Herb. No. 8: 20 (1969)

Small perennial with filiform long-creeping much-branched rhizomes, forming [99] loose mats or dense short turf; *culms* filiform, usually 1–5(–10) cm long, longer or shorter than the leaves; *leaves* 1–4, filiform, obtuse, channelled, ±0.5 mm wide; *sheaths* finely nerved, hyaline, pale brown or ± tinged with red, truncate at the orifice; *spikelets* 1–2(–3), ovoid, the subtending bract erect, leaf-like, ± 1–3 cm long; *glumes* membranous, 0.8–1.3 mm long, nearly as wide as long, blunt, emarginate or the green keel very shortly excurrent, the sides finely nerved and ± stained with red, the margins hyaline; *stamen* 1, occasionally 2–3 in the lowermost florets; *anthers* 0.25–0.5 mm long; *style branches* 3; *nut* ± 0.7–1(–1.1) × 0.5–0.7 mm, subequal to or slightly shorter than the glume, broadly elliptical in outline, apiculate, subequally trigonous, smooth, pale stramineous, the angles prominently ribbed.

Tablelands and ranges from SE. Queensland to Tasmania, also mountains of *Distribution* New Zealand (North Island), New Guinea, the Philippine Islands, Celebes and North Borneo (Mt Kinabalu).

Damp areas in short alpine herbfield and near the margins of lakes, streams *Notes and* and bogs. *Habitat*

References: Blake (1947); Kern (1974).

S. *habrus* Edgar, N.Z. J. Bot. 4: 199 (1966)

Densely tufted perennial from ascending or shortly creeping rhizomes; *culms* [100] very slender, (1.5–)10–20 cm or more high, less than 0.5 mm diam.; *leaves* usually 1–3 per culm, less than 0.5 mm diam., shorter than the culms; *sheaths* usually tinged with red-purple; *spikelets* 1–3, 2.5–4 mm long, the subtending bract 0.5–2.5 cm long; *glumes* 1–2 mm long, ovate-elliptic, acute, the keel green and usually shortly excurrent, the sides usually strongly tinged with red-purple; *stamens* 3–2 in the lower florets, usually 1 in the upper florets; *anthers* 0.5–0.7 mm long; *nut* (0.9–)1–1.4 × 0.6–0.8 mm, obovate-elliptic to broadly elliptic in outline, mucronate, trigonous with the angles slightly thickened.

Tablelands and ranges of eastern N.S.W. and eastern Victoria, mountains of *Distribution* Tasmania and New Zealand, and Stewart, Chatham, Auckland and Campbell Islands.

This species is usually larger in all its parts than *S. subtilissimus,* but dwarf plants *Notes and* from high altitudes are sometimes very difficult to distinguish from the latter. *Habitat* The two species are sometimes found growing in close proximity, although *S. habrus* appears to prefer slightly wetter and more sheltered situations.

S. *montivagus* S. T. Blake, Proc. R. Soc. Queensl. 60: 46 (1949)

Densely tufted perennial forming dense leafy cushions or short swards from as- [101] cending or shortly creeping rhizomes; *culms* very short, crowded, 0.5–2(–4) cm high, shorter than the leaves and bracts; *spikelets* 1–3, ± hidden among the foliage, subtended by 1–2 long leaf-like bracts; *glumes* 1.5–2.2 mm long, ovate-oblong, the keel thickened distally and percurrent or very shortly excurrent, pale green or the sides minutely red-flecked or sometimes narrowly tinged with red-purple towards the hyaline margins; *stamens* 3–2 in the lowermost florets, 1 above; *anthers* 0.6–1.2 mm long; *nut* 1–1.4 × 0.7–0.9 mm, obovate or obovate-elliptic in outline, variably mucronate, the dorsal angle not or only obscurely ribbed, stramineous becoming brown and shining at maturity, shorter than the glume and often falling enclosed in it.

Distribution   Alpine and subalpine tracts of the Australian Alps and associated ranges as far north as the A.C.T.

Notes and Habitat   This species usually forms dense cushions, patches or short swards and is often intermixed with other sward-forming species, e.g. *Scirpus subtilissimus* etc. It is found in short alpine herbfield and on the margins of lakes, streams and bogs. A dwarf form with very small spikelets is found in small depressions in sod tussock grassland.

Carpha Banks & Soland. ex R. Br.

C. nivicola F. Muell., Trans. Phil. Soc. Vict. 1: 111 (1855)
'Broad-leaf flower-rush'

[102]   Loosely tufted glabrous perennial about 10–30 cm high from ascending rhizomes; *leaves* moderately stiff, obtuse, channelled, ± flattened distally, ± 1–3 mm wide, the margins scaberulous; *sheaths* broad, brown or dark brown and shining; *spikelets* 1-flowered, ± 15 mm long, subtended by small glumaceous bracts, arranged in a loose terminal corymb-like cluster subtended by 2(–3) leaf-like bracts; *glumes* (4–)5, lanceolate, distichous, stramineous, keeled above, the two outer small and empty, the two larger glumes subtending the flower, with a linear or setaceous empty glume above; *hypogynous bristles* 6, 13–15 mm long, pale ferruginous, plumose except the tips antrorsely scabrid, falling attached to the nut; *stamens* 3; *style branches* 3; *nut* pale to dark brown, on a short stipe ± 0.5 mm long, the body ± 3.5–4 × 1.2 mm, oblong-elliptic in outline, trigonous, the persistent style-base rigid, terete, ± 6–9 mm long.

Distribution   Alpine and subalpine tracts of the Australian Alps from the Kosciusko area, N.S.W., to Mt. Wellington and the Baw Baws in Victoria.

Notes etc.   Fens, bogs and other wet areas, often in disturbed sites.

Reference: Blake (1940).

C. alpina R. Br., Prodr. Flor. Nov. Holl. 230 (1810)
'Small flower-rush'

Tufted perennial ± 5–25(–30) cm high from short ± erect rhizomes; *leaves* [103]
moderately stiff, usually shorter than the culms, 0.5–1.5(–2) mm wide, channelled, ± flattened distally, obtuse, the margins scaberulous; *sheaths* broad, brown
and shining; *spikelets* 1-flowered ± 10(8–12) mm long, subtended by small
glumaceous bracts, arranged in a loose terminal corymb-like cluster subtended
by 2–3 leaf-like bracts, usually with 1 or more stalked clusters below; *glumes*
(4–)5, distichous, chartaceous, stramineous, lanceolate, subobtuse, keeled above,
the lower 2–3 small and empty, two larger glumes subtending the flower, with
a linear or setaceous empty glume above; *hypogynous bristles* 6, 7–10 mm long,
pale ferruginous, plumose except the tips antrorsely scabrid, falling attached to
the nut; *stamens* 3; *style branches* 3; *nut* pale to dark brown, on a short stipe ±
0.5 mm long, the body oblong in outline, trigonous, ± 2.5–3.3 × 0.8–1 mm,
the persistent style-base rigid, terete, ± 3–5 mm long.

Alpine and subalpine tracts of the Kosciusko area, N.S.W., Mt Hotham, Mt   *Distribution*
Baw Baw and the Bogong High Plains, Vic., and high mountains of Tasmania,
New Zealand and New Guinea.

Usually more erect and delicate in appearance than *C. nivicola* in the Kosciusko   *Notes and*
area, this species is found in similar habitats, being especially associated with small   *Habitat*
depressions in sod tussock grassland and wet disturbed sites in alpine herbfields.
Reference: Kern (1974).

Schoenus L.

S. calyptratus Kükenth., Feddes Repert. 48: 248 (1940)
'Alpine bog-rush'

Small densely tufted or mat-forming glabrous perennial ± 2–5 cm high, from   [104]
slender ascending branched rhizomes; *culms* branched, leafy, up to 2.5 cm long;
leaves much overtopping the culms, stiff, setaceous, ± 2–5 cm × 0.4 mm, straight
or curved, channelled, subacute and usually serrulate at the apex; *sheaths* open,
deeply stained with dark reddish purple, the margins hyaline; *spikelets* usually
1–3 per culm, only shortly exserted from the upper leaf sheath, 1-flowered,
4–5 × 1–1.5 mm, lanceolate in outline, subtended by 2–3 leafy bracts; *glumes*
4, distichous, subacute, keeled above, pallid or very pale green, sometimes tinged
with reddish purple, the 2 lower empty and shorter than the upper; *stamens*
3, the filaments elongating at anthesis; *style* slender, 1.5–2 mm long; *stigmas* filiform; *hypogynous bristles* 5–7, 2–3 mm long, pale, filiform, antrorsely scaberulous,
attached to and falling with the nut; *nut* ellipsoid, 1.2–1.8 × 0.7–0.9 mm, pale
and ± shining, punticulate or almost smooth, 3-ribbed, the ribs thickened and
coalescing at the apex to form a small apiculate cap.

Alpine tract of the Kosciusko area, N.S.W., and the Victorian Alps.   *Distribution*

Common in short alpine herbfield, sometimes forming extensive short turf.   *Notes etc.*
Reference: Blake (1941).

Oreobolus R. Br.

O. distichus F. Muell., Trans. Phil. Soc. Vict. 1: 109 (1855)
'Fan tuft-rush'

[105] Small glabrous perennial with closely packed culms forming dense tufts, cushions or mats 3–8(–12) cm high; *leaves* markedly distichous, rigid, linear-subulate, acute, ± 2–6 cm × 1 mm, incurved or spreading, channelled proximally, ± flattened distally, the margins serrulate, the upper surface obscurely 2-veined with a broad median band between, the underside obscurely 3-veined with the lateral veins not reaching to the apex; *sheaths* broad, equitant, 5–7-nerved, the membranous margins minutely ciliolate, the upper sheaths ± auriculate at the apex; *spikelets* 1-flowered; *peduncles* subterete, elongating at maturity and at length subequal to or usually overtopping the leaves; *glumes* 3, caducous, the inner glumes subequal, 6–8.5 mm long, green with membranous margins, the outer glume bract-like and 6–12 mm long; *stamens* 3; *hypogynous scales* 6, pallid, 1–1.5 mm long; *nut* glabrous, 1.5–2.2 × 1–1.2 mm, ellipsoid to subpyriform, acute or obtuse at the apex, much exceeding the hypogynous scales.

Distribution Alpine and subalpine tracts of the Australian Alps and associated ranges as far north as the A.C.T., and Cradle Mountain, Tas.

Notes etc. Margins of bogs, short alpine herbfield and wet areas in tall alpine herbfield. Reference: Edgar (1964).

O. pumilio R. Br., Prodr. Flor. Nov. Holl. 236 (1810)
'Alpine tuft-rush'

[106–8] Small glabrous perennial up to a few cm high, forming dense cushions, mats, or extensive short swards; *culms* closely packed, covered at the base with the remains of old leaves; *leaves* not markedly distichous, rigid, glabrous, ± 1–2.5 cm × 1–1.5 mm, 5(–6)-veined, channelled proximally, distally flattened and slightly wider with scaberulous margins; *sheaths* broad, imbricate, 6–8-nerved, without conspicuous auricles, the margins membranous and minutely ciliolate; *spikelets* 1-flowered; *peduncles* ± flattened above, elongating at maturity and at length subequal to or shortly overtopping the leaves; *glumes* 3(–4), lanceolate, caducous, the inner acute or acuminate, ± 4–7 mm long with green keels and membranous margins, the outer glume sometimes longer and bract-like with a short rudimentary lamina; *stamens* 3; *anthers* 2.5–3.5 mm long; *hypogynous scales* 6, narrow lanceolate, brown or stramineous, 1.3–1.9 × 0.3 mm, acute, the margins antrorsely scaberulous distally; *nut* pale brown, ± 1.3–1.5 × 1 mm, obovoid-ellipsoid, truncate with a shallow hispidulous depression at the apex, subequal to or shorter than the hypogynous scales.

Alpine and subalpine tracts of the Kosciusko area, N.S.W., Mt Gingera, A.C.T., the Bogong High Plains, Vic., and high mountains of Tasmania and New Guinea. *Distribution*

Short alpine herbfield, margins of fens, and wet areas along the margins of streams. *Notes and* The cushions sometimes die off in the middle and the margins continue to grow *Habitat* outwards forming rings up to about 30 cm diam.; this is a fairly common feature of *Oreobolus* (Kern 1974, p. 441) and also occurs in certain grasses, e.g. *Triodia*, the Spinifex of the inland; mushroom fairy rings are formed in a generally similar fashion.

References: Edgar (1964); Kern (1974).

Uncinia Pers.

U. sinclairii Boott in Hook. f., Handb. N.Z. Fl. 309 (1864)

Rhizomatous perennial to about 15 cm high; *rhizomes* thin, wiry, ± 1–1.5 mm [109] diam.; *culms* subtrigonous, smooth, subequal to or overtopping the leaves; *leaves* short, rather stiff, ± 3–6(–8) cm × 2–4 mm, flat, keeled, acuminate, the margins scaberulous; *spike* dense, ± 1–2 cm × 4–7 mm, oblong or elliptic in outline, brown or stramineous, ebracteate or occasionally the lowermost glume bract-like with a short antrorsely scabrid awn up to 5 mm long, the terminal male portion inconspicuous and ± 8 mm long; *glumes* persistent, coriaceous, ± 5 × 4 mm, broadly ovate, obtuse, the central nerves conspicuous; *utricles* ± 5–6 × 1.5–2 mm, elliptical, plano-convex, brown, finely nerved, hispid on the margins and both surfaces in the upper half, subequal to or usually slightly longer than the glumes.

Distribution   Alpine tract of the Kosciusko area, N.S.W., and New Zealand (South Island).

Notes and   This species was first collected in the Kosciusko area in February 1972. It is at
Habitat   present only known from near the Snowy River Bridge below Seaman's Hut, colonizing bare areas between *Poa* tussocks in sod tussock grassland, and may be an introduction.

References: Gray (1974); Hamlin (1959).

U. sp.

[110]   Densely tufted perennial; *culms* trigonous, stiff, erect, usually shorter than the leaves, ± (5–)10–15 cm long, striate; *leaves* stiff, erect, becoming ± incurved or twisted above, channelled or closely folded, ± 1–1.5 mm wide, attenuate, obscurely scabrid on the margins and distally on the keel; *outer sheaths* reddish brown drying brown; *spike* oblong, ± 1.5–3 cm long incl. the 7.5–14.5 mm long male portion; *glumes* ovate-lanceolate, 5.2–6.5 × ± 2 mm, green on the back, the pale membranous margins sometimes faintly tinged with reddish brown, soon stramineous, the midrib of the lower 1–2 glumes produced into a scabrid awn; *utricles* smooth, brown and shining and obliquely spreading at maturity, (4.2–)5–5.5 × 1.8–2.1 mm, ovate-lanceolate to elliptic in outline, gradually narrowed to the base, acuminate to the oblique orifice, trigonous, with 2(–4) conspicuous nerves; *rachilla* up to about twice as long as the utricle.

Distribution   Alpine tract of the Kosciusko area, N.S.W., and Mt Spion Kopje, Bogong High Plains, Victoria.

Notes etc.   Common in *Celmisia–Poa* tall alpine herbfield and sod tussock grassland.

U. flaccida S. T. Blake, Proc. R. Soc. Queensl. 51: 49 (1940)
'Mountain hook-sedge'

Densely tufted perennial; *culms* very slender, shorter or longer than the leaves, ± (10–)20–40 cm high, acutely trigonous, striate, usually scabrid on the angles distally; *leaves* thin, narrow, grass-like, ± (0.6–)1–2 mm wide, flat or folded, attenuate, slightly keeled, the keel and margins scabrid above the middle; *sheaths* thin, the outer pale brown or reddish brown; *spike* narrow-oblong to oblong, (1–)2–3.5 cm long incl. the 6–11 mm long male portion; *glumes* ovate, ± 4–5 × 1.2–1.5 mm, green on the back, with three main nerves and variable inconspicuous incomplete nerves, the sides hyaline and ± tinged with pale yellowish brown, subequal to or usually slightly shorter than the utricle, the midrib of the lowermost glume produced into a scabrid awn shorter or longer than the spike; *utricles* smooth and shining, obliquely spreading at maturity, (4.5–)5(–5.5) × 1.5–1.8(–2) mm, ovate-lanceolate to elliptic in outline, gradually narrowed to the base, acuminate to the ± oblique orifice, trigonous, with 2(–4) conspicuous nerves; *rachilla* up to about twice as long as the utricle.

Distribution   Alpine and subalpine tracts of the Australian Alps and associated ranges, and mountains of Tasmania.

This species is rare above the treeline and sparsely distributed in the subalpine tract. It has been found in tall alpine herbfield and sod tussock grassland above Club Lake and Lake Albina. *Notes and Habitat*

U. compacta R. Br., Prodr. Flor. Nov. Holl. 241 (1810)

Close to the preceding (*U. flaccida*), differing mainly as follows: loosely tufted; *leaves* slightly thicker and stiffer, shorter and broader, ± 2–3.5 mm wide, often broadly keeled, the margins usually more conspicuously scabrid; *sheaths* usually darker, dark reddish brown; *spikes* generally wider and more congested; *glumes* ± 2–2.5 mm wide, darker, tinged with brown or reddish brown on the sides (but soon fading), and the utricles, in the Kosciusko specimens, are generally slightly wider.

Alpine tract of the Kosciusko area, N.S.W., Mt Baw Baw in Victoria, and mountains of Tasmania. *Distribution*

Occasional in tall alpine herbfield and sod tussock grassland. The fruit of this species is very commonly infected by a black smut in the Kosciusko area. *Notes and Habitat*

Carex L.

KEY TO THE SPECIES

1	Spike solitary, the male flowers above and female below; margins of utricles smooth	*C. cephalotes* (p. 146)
1	Spikes more than one (occasionally solitary in depauperate *C. hebes*, but then the female flowers above the male and the utricles setulose on the upper margins)	2
2	Stigmas 3; culms short, the inflorescence ± hidden among the leaves at the base of the plant	3
2	Stigmas 2; inflorescence well exserted at maturity	4
3	Utricles pubescent, 2.5–3.5 × 1–1.5 mm	*C. breviculmis* (p. 146)
3	Utricles glabrous, 4–5.7 × 1.7–2.5 mm	*C. jackiana* (p. 146)
4	Terminal spike (or spikes) wholly male	*C. gaudichaudiana* (p. 147)
4	Terminal spike with female flowers above and male below	5
5	Glumes blackish on the sides	*C. hypandra* (p. 147)
5	Glumes not blackish on the sides	6
6	Utricles reflexed at maturity, the prominent beak subequal in length to the body	*C. echinata* (p. 148)
6	Utricles not reflexed, the beak shorter than the body	7
7	Glumes pale and inconspicuous, hyaline, very faintly tinged with pale yellowish brown on the sides; utricles ± 2–3 × 1–1.5 mm	*C. curta* (p. 148)
7	Glumes conspicuous, reddish brown or castaneous on the sides with hyaline margins; utricles 3–3.7 × 1.5–1.7 mm	*C. hebes* (p. 149)

C. cephalotes F. Muell., Trans. Phil. Soc. Vict. 1: 110 (1855)

[111]   Glabrous densely tufted perennial about (5–)15–20(–25) cm high from ascending rhizomes; *culms* smooth, terete; *leaves* usually shorter than the mature culms, channelled or ± involute, gradually tapering to a fine keeled tip, antrorsely scaberulous on the margins; *spike* solitary terminal, androgynous, ± (5–)7–15 × 5–9 mm, at first ovoid with the male flowers almost hidden, the utricles spreading or reflexed at maturity, and then the female portion of spike cylindrical and longer than the slender distal male portion, usually ebracteate but sometimes a well-developed bract present; *female glumes* ovate, membranous, deciduous, reddish brown on the sides with paler midrib and hyaline margins, subequal to or shorter than the utricles; *utricles* smooth, usually shortly stipitate, ± 3–4 × 1–1.4 mm, elliptic-lanceolate, pale green proximally becoming tinged with dark reddish brown distally, the orifice of the beak membranous and oblique or ± erose; *stigmas* usually 2.

Distribution   Alpine tract of the Kosciusko area, N.S.W., and the Victorian Alps (Mt Hotham and Mt Bogong); also mountains of New Zealand.

Notes and Habitat   Flowering and fruiting specimens look rather dissimilar for the reasons given above. This species is very closely related to the European *C. pyrenaica* Wahl., and is regarded by some authors as merely a variety of it. Found mainly in wet sites in bogs, fen, short alpine herbfield and sod tussock grassland.

C. breviculmis R. Br., Prodr. Flor. Nov. Holl. 242 (1810)

[112]   Tufted perennial, ± 5–15(–30) cm high from tough woody ascending or shortly creeping rhizomes, the base of the plant often surrounded by the fibrous remains of old sheaths; *culms* usually only a few cm long and hidden among the leaves (elongating in well-grown specimens from lower altitudes); *leaves* overtopping the culms, ± 1.5–3 mm wide, folded, the margins and keel antrorsely scaberulous at least distally; *spikes* 2–5(–7), the terminal spike male; *female spikes* suberect, pale green, ± 7–15(–25) × 3–5 mm, sessile or shortly pedunculate; *glumes* ovate, pale, membranous, the greenish midrib excurrent in a scabrid awn 0.2–2.5(–4) mm long; *utricles* pubescent, pale yellowish green, 2.5–3.5 × 1–1.5 mm incl. the 0.5-mm beak, subequal to or usually shorter than the glumes, broadly fusiform and swollen about the middle, stipitate, finely many-nerved; *stigmas* 3; *nut* brown, broadly obovoid, obtusely trigonous, ± 1.5 × 1.1 mm, the style base dilated and persistent as a minute annulus.

Distribution   Tablelands and ranges of eastern Australia from Queensland to Tasmania, and south-eastern South Australia; also New Zealand (North and South Islands), Lord Howe Island, New Guinea, Java and Celebes.

Notes and Habitat   Fairly common in tall alpine herbfields and sod tussock grassland. This species, like *C. hebes*, occurs on drier sites than most other *Carex* species.

C. jackiana Boott, Proc. Linn. Soc. Lond. 1: 260 (1845)

[113]   Tufted perennial often forming ± dense compacted swards up to 1 m or more in diam.; *culms* very short, the inflorescence ± hidden among the leaf bases;

leaves dark green or bluish green, ± flat or broadly keeled, ± 5–15(–25) × 0.3–0.7 cm, the upper surface with a prominent lateral nerve on either side of the mid-nerve, the margins scabrid; *glumes* deciduous, shorter than the utricles, ± 2–3 mm long, ovate, with a 3-nerved greenish centre and broad whitish membranous margins, acute, acuminate or the nerves coalescing and excurrent into a short awn; *utricles* glabrous, deciduous, stipitate, greenish, 4–5.7 × 1.7–2.5 mm, fusiform-ellipsoid, swollen to subtrigonous, faintly many-nerved, gradually tapering to the long conical beak; *nut* pale brown, stipitate, obovoid-ellipsoid, trigonous, 2–2.5 × 1.3–1.7 mm incl. the short persistent style base which is not dilated; *stigmas* 3.

Alpine and subalpine tracts of the Kosciusko area, N.S.W., and the higher parts of the Victorian Alps; also in Java, Sumatra, Malaya, Ceylon and India.   *Distribution*

As Willis (1970) points out, the Australian form of this species has shorter culms and more congested inflorescences than the tropical form. The glumes and utricles are readily deciduous at maturity. Typical habitats for *C. jackiana* are the margins of bogs, and wet areas in sod tussock grassland.   *Notes and Habitat*

Reference: Willis (1970).

C. gaudichaudiana Kunth, Enum. Pl. 2: 417 (1837)

Glabrous tufted perennial ± 20–40(–80) cm high from tough ascending or long-creeping rhizomes; *culms* trigonous, the angles smooth or antrorsely scabrid; *leaves* dark green, flat, keeled, ± 2–4 mm wide, the margins antrorsely scabrid; *spikes* usually 3–5, the lower subtended by leaf-like bracts, the uppermost spike wholly male, sometimes subtended by 1–2 smaller male spikes, the remainder female or variably male at the top; *glumes* ovate, obtuse or subacute, dark reddish brown or usually almost black on the sides, the greenish midrib not or scarcely reaching the apex; *utricles* ± 3–4 × 1.2–1.6 mm, usually exceeding the glumes, pale green, shortly stipitate, compressed, ovate to elliptic in outline, gradually narrowed to the short beak, strongly 4–6-nerved on both faces; *stigmas* 2.   [114–16]

Tablelands and mountains of south-eastern Australia (incl. Tasmania), New Zealand and New Guinea.   *Distribution*

A main dominant in fen and some bogs, and also common in wet areas of sod tussock grassland. *C. gaudichaudiana* forms compact peats, some of which have been dated to help establish the more recent climatic history of the Kosciusko area. The oldest peats dated, on glacial or periglacial surfaces, are about 15 000 years old, indicating that glacial ice and snow had by then disappeared sufficiently for plant growth to commence.   *Notes and Habitat*

Reference: Costin (1972).

C. hypandra F. Muell. ex Benth., Flor. Aust. 7: 439 (1878)

Very close to high-altitude forms of *C. gaudichaudiana* and virtually indistinguishable vegetatively, differing in the thicker, more congested inflorescence ± 2–4(–6) cm long, the shorter, thicker spikes, the terminal one of which is   [117] female in the upper part, and the broadly ovate, scarcely beaked utricles, ± 2.4–3.3 × 1.4–2 mm.

Distribution   Alpine tract of the Kosciusko area, N.S.W., and Cradle Mountain, Tas. (Willis 1956*a*).

Notes and   This species is found in fens and bogs, especially at higher levels, and is sometimes
Habitat   found growing with *C. gaudichaudiana*.
Reference: Willis (1956*a*).

C. echinata Murr., Prodr. Stirp. Götting. 76 (1770)   'Star sedge'

[118]   Slender tufted glabrous perennial to about 15 cm high from short ascending rhizomes; *culms* slender, trigonous, striate, smooth or scarcely scaberulous just below the inflorescence; *leaves* 0.5–1.5 mm wide, flat or folded, keeled, antrorsely scabrid on the margins, gradually tapering to a slender ± trigonous tip; *inflorescence* pale brown or greenish, ± 1–3 cm long, subtended by an inconspicuous glume-like or setaceous bract, or sometimes a conspicuous foliaceous bract developed; *spikes* 3–5, sessile, ± 3–7 mm long, the female flowers above and male flowers at the base, the lower spikes appearing wholly female, each spike usually developing only a few utricles; *glumes* ovate, acute, or the midrib minutely excurrent, pale brown with greenish midrib and broad hyaline margins, about half the length of the utricle; *utricles* sessile, spreading or strongly reflexed at maturity, ± 3–3.5 × 1–1.5 mm, the body ovate in outline, plano-convex, subtruncate at the base, gradually tapering above to the prominent flattened beak which is about as long as the body, setulose on the margins, and slightly bifid at the orifice; *stigmas* 2.

Distribution   Alpine and subalpine tracts of the Australian Alps; also New Zealand, northern Europe and Asia and North America.

Notes and   Uncommon in short alpine herbfield, and in moist sites in tall alpine herbfields.
Habitat   Willis (1970) suggests that the Australian populations of this widespread species may warrant varietal rank. The star-shaped arrangement of the reflexed utricles, only a few of which are developed in each spike, make Star Sedge easy to recognize.
Reference: Willis (1970).

C. curta Gooden., Trans. Linn. Soc. Lond. 2: 145 (1794)

[119]   Glabrous loosely tufted perennial about 10–40 cm high from ascending or shortly creeping rhizomes; *culms* acutely trigonous, antrorsely scaberulous below the inflorescence, subequal to or usually overtopping the leaves; *leaves* pale green, flat or folded, keeled, ± 10–20 cm × 1.5–4.5 mm, gradually tapering to a fine point, the margins and keel antrorsely scaberulous distally; *inflorescence* pale yellowish green, ± 1.5–3 cm long, the bracts usually glume-like and inconspicuous, or sometimes the lowermost spike subtended by a setaceous or sub-foliaceous bract; *spikes* 3–8, up to 1 cm long, wholly female in appearance but most with a short male portion at the base; *glumes* ovate, acute or shortly acuminate, shorter than the utricles, hyaline, very faintly tinged with pale yellowish brown, the midrib greenish; *utricles* sessile or stipitate, 2–3 × 1–1.5 mm incl. the short notched

beak, ovate or ovate-elliptic in outline, plano-convex, strongly nerved on both faces, ± setulose on the upper margins; *stigmas* 2.

Alpine and subalpine tracts of the Australian Alps. This is a widespread species of mountainous areas in Europe, Asia, New Guinea, North and South America, Fuegia and Falkland Islands. *Distribution*

This species is readily recognized by its pale yellowish green inflorescences. It occurs in fens and bogs, and wet sites in sod tussock grassland and tall alpine herbfields. For a comparison between northern and southern hemisphere populations, see Moore and Chater (1971), who suggest that the predominantly southern hemisphere form with wider stems and leaves and larger utricles should be referred to var. *robustior* (Blytt ex Andersson) Moore & Chater. *Notes and Habitat*

Reference: Moore and Chater (1971).

C. hebes Nelmes, Kew Bull. 1939: 310 (1939)

Tufted perennial ± 10–20(–30) cm high with ascending or shortly creeping tough woody rhizomes ± 2–3 mm in diam; *culms* acutely trigonous, the angles scabrid below the inflorescence; *leaves* flat or channelled, convolute when dry, 1–2 mm wide, usually shorter than the culms; *inflorescence* congested, usually shorter than the subtending setaceous or subfoliaceous bract; *spikes* 3–5, ovoid or ellipsoid, ± 5–11 mm long, wholly female in appearance, with a few inconspicuous male flowers at the base; *glumes* ovate, acute or subacute, subequal to or shorter than the utricles, reddish brown or castaneous on the sides, with a broad pale or green midrib and hyaline margins; *utricles* plano-convex, substipitate, 3–3.7 × 1.5–1.7 mm including the short flattened beak, the body scarcely nerved, ovate-elliptical in outline, narrowly winged with setulose margins above; *stigmas* 2. [120]

Alpine and subalpine tracts of the Australian Alps and associated ranges as far north as the Brindabella Range, A.C.T. *Distribution*

A widespread sedge in several communities, especially in drier sites in tall alpine herbfields and sod tussock grassland (cf. *C. breviculmis*). *Notes and Habitat*

Restionaceae

Empodisma Johnson & Cutler

E. minus (Hook. f.) Johnson & Cutler, Kew Bull. 28: 383 (1973)
'Spreading rope-rush'

Dioecious rhizomatous perennial; *culms* green and glabrous, thin and wiry, much-branched, ± flexuose, about 15–50 cm × 0.5–1 mm (up to 200 cm long at lower altitudes), terete or grooved on one side, often growing through *Sphagnum*; *leaf sheaths* and *floral bracts* 3–10 mm long, appressed, split to the base with overlap- [121]

ping margins, with a rudimentary spreading or reflexed subulate lamina 1–4 mm long; *spikelets* axillary, ± hidden in the uppermost sheaths; *prophyll* 3–4 mm long, blunt, 2-keeled, the keels with long crinkled septate hairs which frequently protrude through the orifice of the sheath or bract; *male spikelets* few-flowered, solitary and sessile, or paired with one pedicellate; *glumes* acuminate or mucronate, usually slightly longer than the flowers; *male flowers: tepals* 6, membranous, ± 3 mm long, narrow-linear, subacute; *stamens* 3; *anthers* versatile, 1.5–2 mm long, shortly exserted on slender filaments; *female spikelets* solitary, 1-flowered; *glumes* 1–3; *female flowers: tepals* 6–4, broadly ovate, membranous, close around the ovary; *styles* 3 or 2, exserted; *nut* smooth, ± 1.5 mm diam., ovoid or ellipsoid, enclosed by the persistent tepals; $2n = 24$ (as *Calorophus minor*)*.

Distribution	Coast, tablelands and ranges of south-eastern Australia from south-eastern Queensland to south-eastern South Australia, Tasmania and New Zealand.
Notes and Habitat	A common and important component of bogs; also in wet areas of sod tussock grassland and tall alpine herbfields. This variable species is usually larger at lower altitudes.

References: Briggs (1966)*; Johnson and Evans (1966).

Juncaceae

KEY TO THE GENERA	1	Leaves and bracts glabrous; capsule many-seeded	*Juncus* (p. 150)
	1	Leaves and bracts hairy on the margins; capsule 3-seeded	*Luzula* (p. 152)

Juncus L.

KEY TO THE SPECIES	1	Leaves flat or channelled, not septate	2
	1	Leaves terete with internal transverse septa	3

2   Dwarf densely tufted mat- or cushion-forming plant to about 3 cm high; flowers few in a solitary cluster which is ± 3–7 mm diam.   *J. antarcticus* (p. 151)

2   Culms usually ± 10–25 cm high from long-creeping branched rhizomes; flowers numerous, in 1–3 dense globular clusters each about 1–1.5 cm diam.   *J. falcatus* (p. 151)

3   Inflorescence of 1(–2) small clusters, each of 1–3 pale green flowers; capsule pale golden brown at maturity; small slender plants, the culms up to ± 3 cm tall   *J.* sp. (p. 151)

3   Inflorescence of usually more than 6 clusters, each of 4–8 dark flowers; capsule dark brown to almost black at maturity; more robust than the preceding, rare above treeline in wet disturbed sites   Jointed Rush *J. articulatus*

J. antarcticus Hook. f., Flor. Antarct. 1: 79, t. 46 (1844)
'Cushion rush'

Dwarf densely tufted perennial usually growing as cushions or sparse mats among mosses; *culms* terete, usually 1–3 cm high, subequal to or shortly overtopping the leaves; *leaves* basal, stiff, ± 1.5–3 cm × 1 mm, linear-subulate, channelled, subterete towards the acute tip; *sheaths* broad with scarious margins; *inflorescence* a solitary terminal 1–4-flowered cluster about 3–7 mm diam; *flowers* 3–3.5 mm long; *tepals* subequal, acute, dark reddish brown or almost black towards the apex; *stamens* 3; *anthers* ± 0.8 mm long; *capsule* brown, shining, 2.5–3 mm long incl. the short mucro, subequal to or shorter than the tepals; *seeds* reddish, 0.4–0.5 × 0.3–0.4 mm, broadly ellipsoid or subglobular. [122–3]

Alpine and subalpine tracts of the Kosciusko area, N.S.W., and the Bogong High Plains, Vic.; also New Zealand (North and South Islands), Stewart, Auckland and Campbell Islands. *Distribution*

Locally common among mosses in short alpine herbfield, and on the margins of bogs and other wet habitats. *Notes and Habitat*

J. falcatus E. Mey., Synops. Luzul. 34 (1823)   'Sickle-leaf rush'

Stoloniferous perennial; *culms* 10–25(–40) cm high, the tufts rather distant on the tough long-creeping branched rhizomes which are 2–3 mm in diam.; *leaves* mostly basal, often with a solitary cauline leaf about the middle of the culm, flat, slightly falcate, sharp-pointed, up to about 3 mm wide, usually shorter than the mature culms; *flowers* in 1–3 dense brown or blackish globular clusters which are each about 1–1.5 cm in diam.; *tepals* ± 4 mm long, finely roughened on the outside, dark reddish brown with a broad greenish midrib and narrow hyaline margins; *stamens* 6; *anthers* ± 1.3 mm long, on short filaments; *style branches* 3, conspicuously fimbriate, pale mauve and very prominent at anthesis; *capsule* dark reddish brown or blackish, shining, subequal in length to the tepals, the short mucro usually in a slight depression at the apex. [124–5]

Alpine, subalpine and montane tracts of south-eastern Australia from the A.C.T. to Tasmania; this species (in the broad sense) is also found in Japan, the Aleutian Islands and western North America. *Distribution*

Locally common in wet mud in fen and bog. *Notes etc.*

J. sp.

Mat-forming perennial with short ascending rhizomes; *culms* tufted, erect, ± 0.5–3 cm tall and about 0.5 mm diam., terete; *leaves* unitubulose, terete, 0.5–1 mm diam., to 5 cm long, usually much exceeding the inflorescence, the auricles ± acute, ± 0.5 mm long; *inflorescence* of 1 head, rarely with a second head, the heads 1–3-flowered, the subtending bract longer than the inflorescence, with a broad scarious base; *tepals* 1.8–2.7 mm long, the outer tepals shorter than or equalling the inner tepals, usually obtuse, stramineous, occasionally reddish, with wide hyaline margins, the outer tepals ovate, the inner narrow lanceolate; *stamens* [126]

6; *anthers* 0.5–0.8 mm long, about half the length of the filaments; *capsule* ovoid, acuminate, pale golden brown, slightly exceeding the tepals; *seeds* ellipsoid, ± 0.5–0.7 mm long.

Distribution   Alpine and subalpine tracts of the Kosciusko area, N.S.W., as far north as the Kiandra district.

Notes and Habitat   Wet situations in short alpine herbfield and near the margins of bogs, streams, etc. This species is superficially very similar to the more widespread *J. sandwithii* Lourteig, which also occurs in the Kosciusko subalpine tract.

References: The above description was kindly provided by Dr L. A. S. Johnson and Mrs K. Wilson of the N.S.W. National Herbarium. The species will be validly published by these authors in a forthcoming revision of Australasian *Juncus*; Lourteig (1968).

Luzula DC.

L. acutifolia Nordenskiöld subsp. nana Edgar, N.Z. J. Bot. 13: 793 (1975)

Dwarf perennial ± 2–5(–8) cm high from short ascending rhizomes, forming mats or extensive swards; *culms* subequal to or shortly overtopping the leaves; *leaves* rather stiff (1–)2–3(–4) mm wide, channelled distally, the tips acute, ± flattened and widened towards the base, persisting around the base of the plant when dry, the margins smooth or papillose, with very sparse long hairs or almost glabrous; *inflorescence* congested, a single blackish subglobose cluster 0.7–1 × ± 0.7 cm; *flowers* (2–)2.4(–2.7) mm long; *tepals* acuminate, dark brown to blackish with inconspicuous narrow paler brown hyaline margins; *stamens* 6, the anthers (0.6–)0.8(–1.2) mm long; *capsule* red-brown to dark brown, subequal in length to the tepals, the surface smooth; *seeds* 1–1.5 × ± 0.7 mm including the small caruncle which is about one eighth the length of the seed; $2n = 12$ (as *L. acutifolia*)*. [127–8]

Endemic to the alpine tract of the Kosciusko area, N.S.W. *Distribution*

Wet sites in short alpine herbfield. *Notes etc.*

References: Edgar (1975); Nordenskiöld (1969)*.

L. novae-cambriae Gandoger, Bull. Soc. Bot. France 46: 392 (1899)

Coarse, loosely tufted perennial (6–)12–20(–30) cm high, the base of the plant usually surrounded by the fibrous remains of old leaves; *culms* much overtopping the leaves in fruit; *leaves* dark green, soft, flat, ± 2–4(–6) mm wide, with or without minute marginal papillae, densely fringed with long white hairs towards the base, sparsely so above, with obtuse-callous tips; *inflorescence* subumbelloid, of several dark brown clusters on peduncles up to 1.5(–2) cm long; *bracts* 1–2, rigid, leaf-like with obtuse-callous tips; *flowers* (2.5–)2.8(–3.2) mm long; *tepals* red-brown to dark brown with rather narrow but conspicuous whitish hyaline margins; *stamens* 6, the anthers (0.7–)0.9(–1.3) mm long; *capsule* subequal to or slightly shorter than the tepals, red-brown to usually dark brown, papillose on the valve margins near the top or smooth; *seeds* 1.3–1.8 × 0.7–0.9 mm including the rather conspicuous caruncle which is about one fifth the length of the seed; $2n = 12$ (as *L. oldfieldii* var. *angustifolia*)*. [129]

Alpine and subalpine tracts of the Australian Alps and also Cradle Mountain, Tas. *Distribution*

Common in rocky areas in tall alpine herbfields and in heaths, this usually robust species is easily distinguished from others above the treeline by the large subumbelloid inflorescence. Edgar (1975) indicates that possible hybrids and hybrid segregates between this species and *L. australasica* appear to be fairly common and may be distinguished by the rather intermediate inflorescence, which is less branched than that of *L. novae-cambriae* but always shows at least some degree of branching. *Notes and Habitat*

References: Edgar (1975); Nordenskiöld (1969)*.

L. atrata Edgar, N.Z. J. Bot. 13: 794 (1975)

[130]   Slender, loosely tufted, rhizomatous perennial about 10–25(–30) cm high; *culms* rigid, erect, much overtopping the leaves at maturity; *leaves* bright green, channelled, (1–)2(–3) mm wide, the margins minutely papillose, fringed with long pale hairs towards the base, very sparsely so above, with obtuse-callous tips; *inflorescence* congested, a single globose blackish head ± 0.7–1 × 0.7–1 cm, subtended by 1–2 acute-tipped bracts with minutely papillose margins, the lowermost bract often long and ± erect; *flowers* (2.2–)2.4(–2.8) mm long; *tepals* dark reddish brown to blackish with quite inconspicuous narrow hyaline margins, acuminate; *stamens* 6, the anthers ± 0.5–0.6 mm long; *capsule* dark reddish brown to blackish, minutely papillose towards the tip, subequal to or slightly shorter than the tepals; *seeds* 1.2–1.4 × 0.6–0.7 mm including the small caruncle which is about one sixth the length of the seed.

Distribution   Alpine and subalpine tracts of the Kosciusko area, N.S.W., Mt Hotham to Mt Loch in Victoria, and Cradle Mountain, Tas.

Notes etc.   Very common in wet areas, near the margins of bogs and fen, and along streams.

L. australasica Steud., Synops. Pl. Glum. 11: 294 (1855)

[131]   Loosely tufted perennial from extensive horizontal or ascending rhizomes ± 1–3 mm in diam.; *culms* erect or ascending, ± 10–30(–50) cm high, much overtopping the leaves at maturity; *leaves* rather flaccid, ± flat, ± 2–5(–7) mm wide, the margins papillose and rather sparsely fringed with weak long white hairs, the obtuse-callous tips usually large and conspicuous; *inflorescence* erect, 1–2(–3) cm long, of 1–4 sessile ± congested clusters, usually elongated and oblong in outline at maturity, the lowermost subtending bracts usually broad and leaf-like; *flowers* (2–)2.5(–3) mm long; *tepals* equal, the outer acuminate, the inner acute, light brown to red-brown at the centre with broad pale hyaline margins; *stamens* 6, the anthers (0.5–)0.7(–0.9) mm long; *capsule* reddish brown, subequal to or usually shorter than the tepals, minutely papillose distally; *seeds* ± 1.2–1.5 × 0.7–0.9 mm including the prominent caruncle which is about one quarter to almost one third the length of the seed; $2n = 24^*$.

Distribution   Alpine and subalpine tracts of the Australian Alps and associated ranges as far north as the A.C.T., with a few outliers as far north as the Barrington Tops, N.S.W., and Tasmania.

Notes and   This species is believed to be pseudo-tetraploid, the chromosome number of
Habitat   24 being derived by each of the original 12 chromosomes breaking in two (Nordenskiöld 1971). It is usually associated with bogs, in which the long rhizomes typically are found growing through the *Sphagnum* moss. See note under *L. novae-cambriae* for hybrids.

References: Edgar (1975); Nordenskiöld (1969*, 1971).

L. oldfieldii Hook. f. subsp. *dura* Edgar, N.Z. J. Bot. 13: 789 (1975)

Coarse tufted perennial (3–)10–20 cm high, usually with persistent dried brown remains of old leaves around the base; *culms* stiff, erect, at first subequal to, but usually much exceeding the leaves in fruit; *leaves* stiff, channelled ± 3–5 mm wide, long-tapering to a small obtuse-callous tip, usually pale green or yellowish green, the nerve-like margins slightly thickened, non-papillose, with very sparse long pale hairs or almost glabrous; *inflorescence* congested into a single globose or conical head ± 1–2 × 1–2 cm, subtended by 2–3 stiff, erect or spreading obtuse-tipped bracts; *flowers* (2.5–)2.7–3.2(–3.5) mm long; *tepals* acute or shortly acuminate, brown or reddish brown with broad whitish hyaline margins; *stamens* 6, the anthers 1–1.5 mm long; *capsule* smooth, red-brown to very dark brown, shorter than the tepals; *seeds* 1.2–1.5 × 0.6–0.8 mm, including the small caruncle which is about one tenth the length of the seed; $2n = 12$ (as *L. oldfieldii* var. *oldfieldii*)*. [132]

Alpine tract of the Kosciusko area, N.S.W., and the Bogong High Plains area, Victoria. *Distribution*

This species and *L. novae-cambriae* are the two largest alpine luzulas. It is recognized by the rather broad, very stiff leaves and large congested heads and is common in exposed sites, e.g. in feldmarks, on large rocky outcrops and on erosion pavements where it is a vigorous colonizer. The Australian subspecies differs from the Tasmanian subsp. *oldfieldii* mainly in the stiff channelled leaves with smooth, scarcely hairy margins, those of the Tasmanian plant being wider and flatter with very hairy and distinctly papillose margins (Edgar 1975). Nordenskiöld (1971) in her hybridization experiments between Australian and New Zealand taxa points out that 'the most fertile hybrids are those between the Australian *L. oldfieldii* and New Zealand taxa'. *Notes and Habitat*

References: Edgar (1975); Nordenskiöld (1969*, 1971).

L. alpestris Nordenskiöld, Bot. Not. 122: 84 (1969)

Densely tufted perennial with short stiff culms about 3–15(–20) cm high; *leaves* stiff, channelled, ± 4–8 cm × (1–)1.4(–2) mm, often curving outwards, with obtuse-callous tips, the pale, slightly thickened nerve-like margins minutely papillose, with sparse, long white hairs above and a dense fringe of hairs near the base of the lamina; *inflorescence* condensed to a single ± ovate to subconical head ± 0.7–1 × 0.5–1(–1.5) cm, occasionally elongated and ± oblong; *bracts* 2–3(–4), stiff, the lowermost frequently ± horizontal or reflexed at maturity, obtuse-tipped or the smaller bracts sometimes sharp-pointed; *flowers* (1.9–)2.1(–2.4) mm long; *tepals* equal, shortly acuminate or the inner acute, pale brown to red-brown at the centre with broad whitish hyaline margins; *stamens* 6, the anthers ± 0.5 mm long; *capsule* reddish brown, usually shorter than or sometimes subequal to the tepals, the valves minutely papillose on the margins near the top; *seeds* 1–1.2 × 0.5–0.7 mm, including the small caruncle which is ± one eighth the length of the seed. [133]

Alpine and subalpine tracts of the Kosciusko area, N.S.W., and the Bogong High Plains, Vic. *Distribution*

Notes and Habitat   Sod tussock grassland and occasionally in wet areas in alpine herbfields. The very dense tufts of short, stiff narrow leaves with a dense fringe of hairs towards the base, the short stiff culms and the relatively pale heads with very small capsules distinguish this species in the field. The leaves of dried specimens are usually greyish green.

References: Edgar (1975); Nordenskiöld (1969).

Liliaceae

KEY   1 Leaves in a dense rosette, with silvery appressed scale-like indumentum at least on the underside; dioecious herbs with small reduced panicles ± hidden among the leaves or only shortly exserted   *Astelia* (p. 157)
1 Leaves glabrous, ± distichous   2

2 Leaves 1–2 cm wide; coarse plants with numerous flowers in panicles
Dianella tasmanica (p. 156)
2 Leaves 2–4 mm wide; small plants with solitary flower almost sessile at the base of the plant   *Herpolirion novae-zelandiae* (p. 157)

Dianella Lam. ex Juss.

D. tasmanica Hook. f., Flor. Tas. 2: 57, t. 133 (1858)
'Tasman flax-lily'

[134–6]   Coarse glabrous loosely tufted perennial about 40–60(–100+) cm high from short ascending rhizomes; *leaves* distichous, crowded towards the base of the stems, ± flat with recurved margins becoming revolute when dry, keeled, ± 30–60 × 1–3(–4) cm, dark green above, pale and slightly glaucous on the underside, serrulate on the margins, coarsely scabrid on the keel; *sheaths* equitant at the base, very coarsely scabrid on the keels above; *panicle* narrow, ± 30–50 cm or more long, the primary branches subtended by boat-shaped bracts which decrease in size up the stem; *flowers* violet-blue, ± 1.5–2 cm diam. on slender ± recurved pedicels; *tepals* 6, subequal, free, usually 5-nerved, the inner with wide membranous margins; *stamens* 6; *anthers* yellow, subequal to or shorter than the swollen orange upper part of the filament; *berry* bright bluish violet at maturity, ovoid-oblong, 1–2 × 1–1.5 cm.

Distribution   Alpine, subalpine and montane tracts of south-eastern N.S.W., as far north as the A.C.T., eastern Victoria, and mountains of Tasmania.

Notes and Habitat   Although common in the subalpine and montane tracts, *D. tasmanica* is rare above treeline, growing in the shelter of rocks above Lake Albina. The blue flowers and brightly coloured berries are very attractive.

Herpolirion Hook. f.

H. novae-zelandiae Hook. f., Flor. N.Z. 1: 258 (1853)   'Sky lily'

Small glabrous tufted perennial forming patches or extensive short grass-like
swards; *rhizomes* wiry, branched, creeping, 1–2 mm diam.; *leaves* linear, distichous,
stiff, bluish green, 2–6 cm × 1.5–3 mm, folded, keeled towards the base, the
outer leaves spreading or ± recurved, the inner short and bract-like; *sheaths*
membranous, open, ± 1–2 cm long; *flower* solitary, pale lilac-blue to almost
white, almost sessile at the base of the plant; *tepals* 6, subequal, free, ± 10–15
× 4 mm, narrowly obovate to oblanceolate, spreading from near the middle;
stamens 6, about half as long as the tepals, the filaments puberulent except at
the apex; *anthers* almost basifixed, 2–3 mm long; *style* undivided, narrowed to
the small stigmatic tip; *capsule* loculicidal, subglobose, obscurely trigonous, slightly
fleshy when fresh, ± 4 mm diam. when dry; *seeds* oblong, smooth, black, shining,
2–3 mm long.

[137]

Alpine and subalpine tracts of the Australian Alps, and mountains of Tasmania
and New Zealand.

Distribution

An uncommon species above treeline in sod tussock grassland, but more common
in the subalpine tract. The flowers are relatively large for such a small plant
and, although attractive, are short-lived.

Notes and Habitat

Astelia Banks & Soland. ex R. Br.

1 Fruit 1-celled, the seeds without flattened faces; male inflorescence exserted, the flowers ± 5–8 mm diam. on slender branches; leaves narrow, acute, ± 0.5–1.5(–2) cm wide *A. alpina* (p. 157)	KEY TO THE SPECIES
1 Fruit 3-celled, the seeds subangular with a few flattened faces; male inflorescence contracted, the flowers ± 1.5–2 cm diam., on stout angular branches; a coarser plant than the preceding with broad attenuate leaves ± 1.5–3(–4) cm wide *A. psychrocharis* (p. 158)	

A. alpina R. Br., Prodr. Flor. Nov. Holl. 291 (1810)
'Pineapple grass; silver astelia'

Dioecious tufted perennial herb forming dense compacted swards on the margins
of *Sphagnum* bogs etc.; *leaves* stiff, crowded in a dense rosette, lanceolate, acute,
mostly 5–15 × 0.5–1.5(–2) cm, ±glabrescent on the upper surface, silvery or
pale ferruginous with dense appressed flattened scale-like indumentum and
strongly keeled on the underside, the broad sheathing bases densely covered
with long white silky hairs; *male inflorescence* loose, exserted, ± 4–8 cm long,
the flowers pale yellowish green, ± 5–8 mm diam., on slender branches; *stamens*
6; *anthers* 0.5–0.8 mm long; *female inflorescence* condensed and ± hidden among

[138–9]

the leaf bases, with fewer flowers on stouter branches; *fruit* a fleshy 1-celled ovoid to oblong berry ± 1 cm × 3–5 mm, yellowish becoming reddish with age; *seeds* obovoid, ± 2 × 1–1.2 mm, smooth black and shining, without flattened faces, about 4–8 per fruit.

Distribution Alpine and subalpine tracts of the Australian Alps, and mountains of Tasmania.

Notes and Habitat The mainland populations generally consist of smaller plants than those from Tasmania, and are separated as var. *novae–hollandiae* Skottsb. (1934). Typical habitats are bogs and wet places in tall alpine herbfields. The coloured berries are particularly attractive. The common name for *Astelia* in Tasmania, Pineapple Grass, aptly describes its rosette of stiff, pointed leaves.

References: Moore, L. B. (1966); Skottsberg (1934).

A. psychrocharis F. Muell., Trans. Proc. Vict. Inst. 1854-55: 135 (1855)

[140–3] Often growing with the preceding species, but usually coarser and more robust, differing mainly as follows: *leaves* usually broader, attenuate at the apex, up to 20 cm long and mostly 1.5–3(–4) cm wide, the sheathing bases less conspicuously silky-hairy; both male and female inflorescences much contracted, the *flowers* larger and usually fewer on very stout angular peduncles; *male flowers* 1.5–2 cm diam. on peduncles ± 1–1.5 cm long, the *anthers* ± 3–3.5 mm long when fresh (drying to ± 1.5–3 mm); *female flowers* almost hidden among the broad spathes and leaf bases, the fruit and seeds larger, the 3-celled *berry* 1.5–2 × 0.5–1 cm (incl. the short beaked ± 2 mm style), obovoid, pale orange-yellow to bright orange when ripe, fleshy proximally, the locules confined to the distal half; *seeds* 2.2–3.3 × 1.3–1.7 mm, black and shining, subangular with a few flattened faces, 4–16 per fruit.

Distribution Endemic to the alpine and subalpine tracts of the Kosciuko area, N.S.W.

Notes and Habitat Common in bogs and wet sites in tall alpine herbfields, often growing with *A. alpina* but in better-drained sites than the latter species. We suggest that an appropriate common name for this species would be Kosciusko Pineapple Grass.

References: Moore, L. B. (1966); Skottsberg (1934).

Orchidaceae

1 Flowers arranged on the stem so that the labellum lies below the column; leaf flat; plant ± hairy towards the base   *Caladenia* (p. 159)

1 Flowers arranged on the stem so that the labellum lies above the column; leaf hollow, tubular; glabrous plants   *Prasophyllum* (p. 159)

Caladenia R. Br.

C. lyallii Hook. f., Flor. N.Z. 1: 247 (1853) 'Mountain caladenia'

Herb ± 10–20(–30) cm high with small fleshy globular tubers; *stem* pilose below   [144]
with spreading septate hairs, glandular-pubescent above; *leaf* solitary, suberect,
thick and slightly fleshy, 6–10(–17) cm × 4–10 mm, elliptic-lanceolate, usually
shorter than the stem, obscurely ribbed, subglabrous or usually with sparse long
septate hairs; *flowers* 1–2(–4) in a lax raceme, the perianth segments glandular-
pubescent and pink or reddish on the back, steely white inside; *dorsal sepal* ovate,
10–14 mm long, arched over the column; *lateral sepals* elliptic-lanceolate, slightly
longer than the dorsal sepal, close together and projecting forward under the
labellum; *petals* spreading, similar to lateral sepals; *labellum* broadly ovate, shortly
clawed, obscurely 3-lobed, transversely red-barred, the mid-lobe triangular, recurved, with toothed margins, the pale or yellow calli arranged in 4 rows, becoming small and ± crowded on the mid-lobe; *column* curved, ± 7 mm long, broadly
winged, blotched or barred with red.

Alpine and subalpine tracts of the Australian Alps and associated ranges as far   *Distribution*
north as the Brindabella Range, A.C.T., with an outlier in the Grampians in
western Victoria, and mountains of Tasmania and New Zealand.

A rare species above treeline, recorded from Mt Northcote, the northern slopes   *Notes and*
of Mt Clarke and the vicinity of Blue Lake. It is more common in the subalpine   *Habitat*
tract, especially around bogs, the preferred habitat.

References: Nicholls (1969); Rupp (1969).

Prasophyllum R. Br.

1	Labellum and petals acute, the petals with obscure narrow pale borders	KEY TO THE
	P. alpinum (p. 159)	SPECIES
1	Labellum and petals obtuse or rounded at the tips, with conspicuous broad	
	white undulate margins	*P. suttonii* (p. 160)

P. alpinum R. Br., Prodr. Flor. Nov. Holl. 318 (1810)
'Alpine leek-orchid'

Usually rather stout glabrous herb ± 10–30(–40) cm high with ovoid fleshy   [145–6]
tubers; *leaf* solitary, sheathing the stem for most of its length, the hollow tubular
lamina overtopping or shorter than the raceme; *raceme* ± 3–10 cm long; *flowers*
green or yellowish green usually ± suffused with red-brown, or sometimes
wholly dark brownish red; *sepals* lanceolate, 4–8 mm long, acute or acuminate,
the lateral sepals usually connate for at least half their length; *petals* acute, linear
or narrowly triangular, subequal to or usually slightly shorter than the sepals;

labellum ± 3.5–5 mm long (when straightened), orbicular proximally, narrowly cuneate distally, abruptly recurved about the middle, the thick callus usually extending almost to the tip.

Distribution   Alpine and subalpine tracts of the Australian Alps and associated ranges as far north as the Brindabella Range, A.C.T., and mountains of Tasmania.

Notes and   Common on the margins of bogs and in seepage areas in tall alpine herbfields
Habitat   and sod tussock grassland. This species shows much variation in flower colour even in the one population. The flowers are fragrant in hot weather. For a note on pollination in this species see Jones (1972).

References: Jones (1972); Nicholls (1969); Rupp (1969).

P. suttonii Rogers & Rees, Proc. R. Soc. Vict. 25: 112 (1912)
'Mauve leek-orchid'

[147]   Vegetatively similar to the preceding, but often larger, about 15–30(–60) cm high, with larger, more attractive flowers; *raceme* 3–11 cm long; *flowers* fragrant, tinged with pink or mauve, the labellum and petals with conspicuous broad, white, undulate margins; *sepals* ± 5–9 mm long, green tinged with reddish purple or mauve, the lateral sepals blunt, usually connate almost to their tips, occasionally almost free but remaining subparallel; *petals* 6–8 mm long, oblong or oblong-spathulate, subobtuse to rounded at the tips, white with a narrow dark or mauve median line, the margins ± undulate; *labellum* very shortly clawed, ± 8 × 4 mm (when straightened), white with conspicuously crisped-undulate margins, oblong to ovate, recurved to form an angle of ± 60° about the middle, obtuse or rounded at the tip, the callus extending to just beyond the bend.

Distribution   Alpine and subalpine tracts of the Australian Alps and associated ranges as far north as the A.C.T., and mountains of Tasmania.

Notes and   Similar habitats to *P. alpinum* and often found associated with that species. It
Habitat   is easily distinguished from *P. alpinum* by the larger, more attractive flowers.

References: Nicholls (1969); Rupp (1969).

ANGIOSPERMAE: Dicotyledoneae

KEY

1 Leaves reduced to small decurrent scales ± 0.5–1 mm long; small creeping subshrub with flattened grooved articulate branchlets and minute flowers; fruit a nut borne on a pedicel which becomes bright red, fleshy and swollen at maturity *Exocarpos nanus* (p. 167)
1 Leaves well developed; plant otherwise not as above 2

2 Individual flowers (florets) small and clustered in dense heads on a common receptacle, surrounded by an involucre of bracts (phyllaries), the whole head resembling a single flower; anthers cohering in a tube around the style; calyx, if present, represented by hairs or scales (pappus); ovary inferior
COMPOSITAE (p. 229)
2 Plants without the above combination of characters 3

3 Leaves mostly hastate with spreading auricles; stipules (ochreae) conspicuous, hyaline, tubular, lacerate; flowers small, numerous in conspicuous reddish panicles; common weedy dioecious herb to 20(–30) cm high, often covering extensive areas Sheep Sorrel *Rumex acetosella* (POLYGONACEAE)
3 Not as above 4

4 Perianth of 1 whorl or of 2 or more similar whorls 5
4 Perianth of 2 (rarely more) distinct whorls, differing markedly from each other in shape, size or colour (in female flowers of *Myriophyllum pedunculatum* the perianth is minute and inconspicuous) 12

5 Perianth petaloid 6
5 Perianth sepaloid 11

6 Stamens numerous; herbs RANUNCULACEAE(in part) (p. 173)
6 Stamens 5 or less, or flowers female; shrubs or herbs 7

7 Leaves crowded, linear, ± 3 mm long, the upper leaves with a conspicuous tuft of white hairs at the tip *Drapetes tasmanicus* (p. 191)
7 Leaves more than 3 mm long, not as above 8

8 Leaves alternate or radical 9
8 Leaves whorled or opposite 10

9 Herbs; ovary inferior, petals and stamens 5; styles 2; fruit of two mericarps
UMBELLIFERAE (in part) (p. 200)
9 Shrubs or subshrubs; ovary superior; perianth segments and stamens 4; style 1; fruit a tough leathery follicle PROTEACEAE (p. 164)

10 Stamens 2; style simple; leaves opposite, exstipulate   *Pimelea* (p. 192)
10 Stamens more than 2 (usually 4); style 2-branched; leaves whorled or opposite
     with interpetiolar stipules   RUBIACEAE (in part) (p. 223)

11 Small glabrous herbs or cushion plants; leaves simple, entire, opposite and
     ± connate at the base   CARYOPHYLLACEAE (in part) (p. 168)
11 Not as above   ROSACEAE (in part) (p. 182)

12 Petals united at least at the base (in *Neopaxia* which has two broad persistent
     sepals, the petals are united only at the very base)   13
12 Petals at length free, or if cohering above, then with claws free at the base
     (e.g. *Stackhousia*), or petals minute or absent (in *Epilobium* the receptacle
     is produced above the ovary to form a very short 'floral tube')   24

13 Corolla zygomorphic (sometimes appearing almost regular in *Veronica* and
     Chionohebe (Scrophulariaceae) but then the stamens 2)   14
13 Corolla regular; stamens never 2   18

14 Ovary superior; stamens 2 (free) or 4 (in 2 pairs with the anthers ± cohering
     by long hairs)   15
14 Ovary inferior; stamens 2 (fused) or 5 (free, or coherent in a tube around
     the style)   16

15 Ovary 4-celled, the fruit separating into 4 nutlets; style basal between the
     4 cells of the ovary; aromatic shrub with conspicuously gland-dotted
     leaves   *Prostanthera cuneata* (p. 215)
15 Ovary 2-celled, the fruit a 2-celled capsule; style terminal; otherwise not
     as above   SCROPHULARIACEAE (p. 216)

16 Anthers free; style with a cup-shaped indusium below the stigma; corolla
     fan-shaped, split to the base on one side, the lobes pubescent on the back
     with broad yellow glabrous wings   *Goodenia hederacea* var. *alpestris* (p. 228)
16 Anthers coherent in a tube around the style, or fused with style to form
     an irritable column; plants otherwise not as above   17

17 Anthers 2, fused with the style to form a geniculate irritable column; erect
     tufted herb with reddish flowers and grass-like leaves
                                          Stylidium graminifolium (p. 228)
17 Anthers 5, coherent in a tube around the style; dwarf prostrate creeping
     herb with pale violet or white flowers and obovate-spathulate to suborbicular
     leaves   *Pratia surrepens* (p. 227)

18 Sepals 2, broad, persistent; prostrate creeping mat-forming herb with white
     or pink flowers   *Neopaxia australasica* (p. 167)
18 Sepals more than 2   19

19 Corolla scarious; scapigerous herbs with flowers few and terminal or in dense
     spikes; capsule circumcissile   *Plantago* (p. 220)
19 Corolla not scarious; plants otherwise not as above   20

20 Corolla campanulate, ± 3–4.5 cm diam., in shades of blue, purple or violet
                                          Wahlenbergia (p. 226)

20 Corolla white, yellow or cream-coloured, or if at length blue then rotate
and ± 2 mm diam. 21

21 Ovary inferior; leaves whorled, or if opposite then with small interpetiolar
stipules RUBIACEAE (in part) (p. 223)
21 Ovary superior; leaves not whorled; stipules absent 22

22 Shrubs or subshrubs with coriaceous leaves EPACRIDACEAE (p. 208)
22 Herbs; leaves soft 23

23 Glabrous herb; leaves opposite; corolla campanulate, shrivelled and persistent
at the base of the fruit; fruit a capsule *Gentianella diemensis* (p. 213)
23 Hairy plant; leaves alternate; corolla rotate, deciduous; fruit of small nutlets
Myosotis (p. 214)

24 Shrub, leaves broadly ovate or oval, the branchlets and underside of the leaves
densely covered with scurfy scales *Phebalium ovatifolium* (p. 188)
24 Herbs or shrubs without scurfy scales 25

25 Carpels and styles quite free or the carpels united only at the extreme base
26
25 Carpels or styles or both united, or carpel 1 28

26 Glabrous dioecious aromatic shrub with rough reddish branches; leaves with
hot peppery taste if chewed; sepals and petals 2 *Tasmannia xerophila* (p. 179)
26 Herbs 27

27 Dwarf annual succulent herb with opposite fleshy leaves; flowers inconspicuous, 4-merous, the petals ± 1 mm long *Crassula sieberana* (p. 181)
27 Perennial herbs with conspicuous flowers; stamens and carpels numerous
RANUNCULACEAE (in part) (p. 173)

28 Perianth (calyx and epicalyx) green and sepaloid; style basally attached on
the single carpel ROSACEAE (in part) (p. 182)
28 Inner perianth whorl petaloid, or flowers female with minute perianth; styles
and carpel not as above 29

29 Perianth zygomorphic 30
29 Perianth regular or almost so 32

30 Shrubs or subshrubs with pea-like flowers PAPILIONACEAE (in part) (p. 183)
30 Herbs . 31

31 Leaves trifoliolate; flowers pea-like; (clovers)
PAPILIONACEAE (in part) (p. 183)
31 Leaves simple *Viola betonicifolia* (p. 190)

32 Petals fused above into a long tube but free at the base; fruit of usually 3
small 1-seeded nutlets; mat-forming plant with fragrant creamy-yellow
flowers *Stackhousia pulvinaris* (p. 189)
32 Petals quite free; plants otherwise not as above 33

33 Shrubs or subshrubs 34
33 Herbs 35

34 Leaves not gland-dotted; flowers yellowish, 2.5–3 mm diam., nodding and
    ± hidden beneath the leaves and branches; anthers 5, sessile, cohering in
    a tube around the ovary; fruit a purplish berry; depressed subshrub with
    spiny branchlets and lenticellate bark   *Hymenanthera dentata* (p. 190)
34 Leaves gland-dotted; flowers erect, conspicuous, the petals white, or if yellow
    then the stamens free, numerous, on long filaments much exceeding the
    petals; fruit a capsule   MYRTACEAE (p. 194)

35 Ovary superior   36
35 Ovary inferior   39

36 Leaves simple, entire   37
36 Leaves lobed or dissected   38

37 Leaves strap-shaped, the margins and upper surface with conspicuous viscid
    glandular hairs; scapigerous insectivorous herb with white flowers
                                                        Drosera arcturi (p. 180)
37 Not as above   CARYOPHYLLACEAE (in part) (p. 168)

38 Leaves palmately lobed or dissected; sepals and petals each 5; stamens 10
                                                            Geranium (p. 186)
38 Leaves pinnately lobed or divided; sepals and petals each 4; stamens 6
                                                        CRUCIFERAE (p. 180)

39 Flowers in umbels and/or the leaves radical; fruit of 2 mericarps
                                            UMBELLIFERAE (in part) (p. 200)
39 Flowers not in umbels, leaves cauline; fruit not as above   40

40 Flowers hermaphrodite, articulate below the calyx, with conspicuous emargi-
    nate petals; style single; fruit a long terete capsule opening from the apex
    by 4 valves; seeds numerous, usually with a tuft of long hairs at the apex
                                                        Epilobium (p. 196)
40 Flowers small, hermaphrodite or unisexual, the petals not emarginate, small
    and hooded in the male and hermaphrodite flowers, absent in the female
    flowers; styles 4; fruit and seeds not as above   HALORAGACEAE (p. 198)

Proteaceae

KEY   1 Perianth glabrous, straight; anthers on short filaments; leaves glabrous
                                                    Orites lancifolia (p. 165)
       1 Perianth appressed-hairy, the limb oblique; anthers sessile; leaves appressed-
          tomentose on the underside   *Grevillea* (p. 165)

Orites R. Br.

O. lancifolia F. Muell., Trans. Phil. Soc. Vict. 1: 108 (1855)
'Alpine orites'

Spreading shrub 1–1.5(–2) m high; *branchlets* and new shoots pubescent with antrorsely appressed pale or ferruginous hairs, soon glabrescent; *leaves* alternate, glabrous, petiolate, 1.5–3 × 0.4–0.8 cm, rigid and coriaceous, oblong-elliptic or oblong-lanceolate, obtuse or subacute, entire or rarely with a few obscure teeth distally, obscurely net-veined and shining above, more prominently veined with slightly thickened margins on the underside; *petioles* 2–5 mm long; *flowers* creamy white, 4–6 mm long in short erect terminal or subterminal spikes 2–5 cm long, the rachis shortly ferruginous-villous; *bracts* deciduous; *perianth* straight, glabrous, 4–6(–8) mm long, the 4 segments splitting to the base, narrowly spathulate; *anthers* ± 0.8 mm long, on short thick filaments attached at the base of the concave laminae of the perianth segments which are reflexed at maturity; *style* straight, 5–7 mm long, glabrous or with a few long hairs towards the base; *ovary* villous; *follicles* tough and leathery, ± 2 cm long, obliquely fusiform with a curved pointed tip, at first grey- or ferruginous-pubescent, glabrescent with age; *seeds* 2, ± 5 mm long with an oblique terminal wing. [148]

Alpine and subalpine tracts of the Australian Alps and associated ranges as far north as the A.C.T. *Distribution*

Fairly common in heaths. *Notes etc.*

Grevillea R. Br. ex Knight

1 Leaves small, rigid, ± 1–2 cm × 1–5 mm; flowers small, creamy-white, 4–6 mm long, in inconspicuous umbel-like racemes shorter than the leaves *G. australis* (p. 165)

1 Leaves large, ± 3–7(–10) × 0.7–2.5 cm; flowers large, dark red or ferruginous, ± 1.5–2 cm long, conspicuous in racemes up to 7 cm long *G. victoriae* (p. 166)

G. australis R. Br., Trans. Linn. Soc. Lond. 10: 171 (1810)
'Alpine grevillea'

Variable shrub, erect, prostrate or scrambling over rocks and boulders; *branchlets* appressed-tomentose, the young shoots and buds ferruginous; *leaves* very variable 1–2 cm × 1–5 mm, coriaceous, rigid, oblong, elliptic, oblanceolate or obovate with strongly recurved margins to linear-revolute, glabrous and often 3–7-nerved above, appressed-tomentose on the underside, shortly mucronate or pungent-pointed; *flowers* 4–6 mm long, strongly scented, creamy white with ferruginous tips, pedicellate in small umbel-like racemes shorter than the leaves; [149–50] *pedicels* 2–4 mm long, ferruginous-pubescent; *perianth segments* appressed-pubescent, glabrous inside (in Kosciusko forms), cohering in bud to form a tube with a recurved globular usually ferruginous limb, splitting to the base and soon deciduous at maturity; *anthers* ± 0.4 mm long, sessile, enclosed in the concave tips of the perianth segments which are reflexed at maturity; *ovary* glabrous, shortly stipitate, subtended by a minute gland; *style* glabrous, recurved distally, with an oblique disc bearing the small stigmatic cone; *follicle* glabrous, broadly ellipsoid or ovoid, 6–12 × 5–6 mm excl. the persistent style.

Distribution	Alpine and subalpine tracts of the Australian Alps and associated ranges as far north as the A.C.T., and mountains of Tasmania.
Notes and Habitat	A widespread component of heaths. This is an extremely polymorphic species, varying considerably in leaf shape; the varieties need further investigation (Curtis 1967; Willis 1973).

References: Curtis (1967); Willis (1973).

G. victoriae F. Muell., Trans. Phil. Soc. Vict. 1: 107 (1855)
'Royal grevillea'

[151–2] Shrub usually ± 1–2 m high with grey- or silvery-tomentose branches, the branchlets and underside of the young leaves pale ferruginous; *leaves* petiolate, subcoriaceous, 3–7(–10) × 0.7–2.5 cm, elliptic, lanceolate or oblanceolate with slightly recurved margins, acute to obtuse, mucronulate, ± pubescent above but soon glabrescent, grey- or silvery appressed-tomentose on the underside; *flowers* red or ferruginous, shortly pedicellate, solitary or paired on the racemes; *racemes* simple or branched towards the base, 1–7 cm long, recurved or reflexed, the rachis appressed-tomentose; *perianth* 1.5–2 cm long, ± densely covered with ferruginous hairs, densely white-bearded inside below the middle, the limb globular in bud; *ovary* glabrous, stipitate, subtended by a small curved truncate gland up to 1 mm long; *style* glabrous, 12–15 mm long, the stigmatic disc oblique; *anthers* ± 1.5 mm long; *follicles* glabrous, ellipsoid, ± 2 × 0.6 cm excluding the persistent style.

Distribution	See Notes.
Notes and Habitat	This species, as it is at present understood, is very polymorphic. The above description applies to the form in the alpine and subalpine tracts of the Australian Alps as far north as Kiandra, in which the limb of the perianth is globular in bud. It is fairly common in the subalpine tract but only very marginal above the treeline, being found e.g. on Sentinel Peak. Another form with the perianth limb angular and pointed in bud and with generally broader leaves and shorter racemes occurs in the A.C.T. and on the ranges to the east thereof.

Santalaceae

Exocarpos Labill.

E. nanus Hook. f., Hook. Lond. J. Bot. 6: 281 (1847)
'Alpine ballart'

Small creeping subshrub with reduced leaves; *stems* creeping and rooting at the [153] nodes, from a few to 25 cm or more long; *branchlets* articulate, grooved and flattened, glabrous, green or yellowish green, procumbent or ascending to 5 cm or occasionally to 10 cm high; *leaves* mostly opposite, reduced to decurrent triangular scales ± 0.5–1 mm long, slightly channelled adaxially, with bluntish dark tips; *flowers* unisexual, small and inconspicuous, ± 1.5 mm diam., 2–3(–4) together in sessile or subsessile axillary clusters, subtended by minute ciliolate bracts; *tepals* (4–)5(–6), broadly triangular, blunt, yellowish green often tinged with red; *stamens* 0.3–0.4 mm long, smaller and sterile in the female flowers; *stigma* flat, sessile; 1–2 fruits developed per inflorescence; *fruiting pedicel* swollen and succulent, bright red and shining, up to ± 5 mm long, obovoid or subcylindrical, crowned by the persistent tepals; *fruit* 2.5–3.5 × 2–2.5 mm, ovoid-ellipsoid to subglobular, green becoming flushed with dark brownish red at maturity, the base sunken in the succulent pedicel, the minute stigma persistent at the apex.

Uncommon and localized in the alpine and subalpine tracts of the Australian *Distribution* Alps and associated ranges as far north as the A.C.T., and mountains of northern Tasmania.

Fairly rare above the treeline but occasionally found growing in heaths and in *Notes and* tall alpine herbfields, usually in the shelter of shrubs and boulders. Like other *Habitat* members of the genus, *E. nanus* is a root parasite, and small haustoria can be seen on the roots. The red fleshy pedicels are edible but not particularly tasty. Reference: Stauffer (1959).

Portulacaceae

Neopaxia Ö. Nilss.

N. australasica (Hook. f.) Ö. Nilss., Grana Palynol. 7: 331 (1967)
'White purslane'

Variable glabrous rhizomatous perennial herb (see Notes below); the alpine [154–7] form usually forms close mats or extensive carpets only a few cm high; *stems* prostrate, creeping and rooting, slightly fleshy, 2–5 mm diam., the old stems with conspicuous leaf scars; *leaves* alternate, erect, 1.5–4 cm long, green or glaucous, linear or narrowly oblanceolate, with a terminal pore (hydathode), expanded and broadly sheathing at the base; *flowers* solitary or in short few–flowered bracteate cymes, white or pink, ± (8–)10–17(–20) mm diam., on slender pedicels 1–2 cm long; *sepals* 2, herbaceous, persistent, suborbicular, 2–4 mm long, imbricate at the base; *petals* 5, obovate, 6–10 mm long, united near the base; *stamens* 5, opposite the petals, the filaments 3.5–5 mm long, dilated at the base; *anthers* extrorse, 1–2 mm long; *style* ± 2 mm long with 3 short stigmatic branches; *ovary* superior; *capsule* subglobular, 3-valved, shorter than the persistent sepals; *seeds* usually 3 (or fewer), smooth black and shining, broadly obovate or suborbicular in outline, 1–1.5 mm diam.; $2n = 96$ (as *Paxia australasica*)*.

Distribution	Widespread in southern temperate Australia, Tasmania, New Zealand (North and South Islands) and Campbell Island.
Notes and Habitat	A characteristic dominant of short alpine herbfield, often forming extensive communities visible as white (and, in places, pink) patches at flowering time, sometimes so dense as to resemble snow from a distance; also present in damp, semi-bare sites, as along stream banks and eroding alpine herbfields. The species is very variable; in addition to the white- and pink-flowered and green- and glaucous-leaved forms found at higher altitudes, there are lower-altitude forms with long weak stems and leaves up to 10 cm long, and forms with fleshy bulbous internodes on the rhizomes. The status of the various forms needs further investigation. Because of its vigorous colonizing ability, *Neopaxia* is being planted out by the Soil Conservation Service of N.S.W. in some of its re-vegetation work on eroded parts of the Main Range.

References: Aston (1973); Nilsson (1966a*, b, 1967).

Caryophyllaceae

KEY	1	Petals absent	2
	1	Petals present	4
	2	Fruit an indehiscent 1-seeded nutlet enclosed by the hardened and swollen perigynous tube *Scleranthus* (p. 169)	
	2	Fruit a capsule opening by teeth or valves; seeds numerous	3
	3	Capsule valves usually as many as sepals or styles; tufted perennial herbs or cushion plants *Colobanthus* (p. 170)	
	3	Capsule valves twice as many as styles; dwarf annual, rare *Stellaria multiflora* (p. 172)	
	4	Stipules conspicuous, scarious; petals pink, entire; capsule 3-valved; annual or biennial herb with slender tap-root and densely fascicled linear leaves; Soil Conservation reclamation areas and disturbed sites Sand-spurrey **Spergularia rubra*	

4   Stipules absent; petals white; capsule 6- or 10-toothed   5

5   Petals deeply bifid, subequal to or exceeding the sepals; capsule cylindrical, 10-toothed, at length curved and much exceeding the sepals; hairy perennial herb, locally common on Soil Conservation reclamation areas and disturbed sites   Mouse-ear Chickweed. *Cerastium fontanum* subsp. *triviale*
5   Petals entire, much shorter than the sepals; capsule flask-shaped, not curved, 6-toothed, shortly exceeding the sepals; minutely pubescent annual, rare casual on Soil Conservation reclamation areas
                    Thyme-leaved Sandwort *Arenaria serpyllifolia*

Scleranthus L.

S. biflorus (J. R. & G. Forst.) Hook. f., Flor. N.Z. 1: 74 (1852)
'Two-flowered knawel'

Much-branched bright green perennial herb, tufted or the stems usually procumbent and forming dense mats or cushions up to a few cm high; *leaves* rigid, crowded in the upper part of the stems, opposite, slightly connate at the base, ± 5–10 mm long, linear-subulate, shortly mucronate, ± serrulate on the margins and sometimes distally on the keel; *flowers* small and inconspicuous in the upper axils, greenish, ± 1–1.5 mm long, sessile in pairs on a glabrous peduncle which elongates up to 2 cm in fruit, one flower subtended by a pair of small ± caducous bracteoles, the pair subtended by two broadly ovate, pointed bracts ± 1 mm long, connate at the base, keeled above and persistent on the top of the peduncle after the fruit have fallen; *sepals* 4–5, ± 0.5 mm long, suberect, persistent, incurved in fruit; *petals* 0; *styles* 2, minute; *stamen* 1; *fruit* an indehiscent 1-seeded nutlet enclosed by the swollen and hardened perigynous tube and the persistent sepals which together are ± 2 × 1 mm.   [158]

Throughout a wide range of altitudes in eastern N.S.W., eastern and southern Victoria, Tasmania and New Zealand.   *Distribution*

Not common above the treeline, but occurs e.g. on damp slopes, near Lake Albina. The compact, mat-forming habit of this species has led to its successful use in rock gardens. '*Scleranthus*' is derived from two Greek words meaning 'hard flower' and refers to the indurated fruiting perianth.   *Notes and Habitat*

S. singuliflorus (F. Muell.) Mattf., Bot. Jb. 69: 272 (1938)
'One-flowered knawel'

[159–60]   Superficially similar to diffuse forms of the preceding, sometimes forming loose clumps or cushions in open places, but usually spreading through grasses with decumbent or ascending stems to 15 cm or more long; otherwise differing mainly in the solitary flowers subtended by 2 transversely oblong truncate bracts with the keel shortly excurrent, the ± scaberulous fruiting peduncles, the leaves ± 4–6 mm long and ± ciliolate on the margins, and the wider fruit (1.7–)2(–2.1) × (1.1–)1.3(–1.5) mm.

Distribution   Alpine and subalpine tracts of the Australian Alps.

Notes and Habitat   Fairly common in tall alpine herbfields and sod tussock grassland; also in *Epacris–Chionohebe* feldmark. Plants from high altitudes on Mt Wilhelm, Mt Strong and The Owen Stanley Range in New Guinea which have been referred to this species, for example by Mattfield (1938; l.c., above), have distinctly narrower fruits and seeds than those from the Australian Alps and need further investigation.

Colobanthus Bartl.

KEY TO THE SPECIES

1	Tufted herb to about 8 cm high; leaves grass-like, 2–6 cm long; fruiting peduncles 2–6 cm long *C. affinis* (p. 170)
1	Compact cushion plants up to a few cm high; leaves <1 cm long; fruiting peduncles to about 1 cm long 2
2	Leaves very stiff, ± spreading, with acicular pungent tips (0.5–)1–1.5 (–2) mm long, the margins and the midrib on the underside thickened and very conspicuous in dried specimens; seeds ± 0.7–0.8 × 0.4–0.5 mm *C. pulvinatus* (p. 171)
2	Close to the preceding, differing mainly in the softer suberect narrower leaves, with generally shorter acicular tips 0.4–0.7 mm long, the margins and the midrib on the underside not or scarcely thickened; seeds ± 0.5–0.6 × 0.3–0.5 mm *C. nivicola* (p. 172)

C. affinis (Hook.) Hook. f., Flor. Tas. 1: 45 (1855)
'Alpine colobanth'

[161]   Glabrous tufted perennial herb 3–8 cm high, usually forming small clumps from a short branched rootstock; *leaves* linear, crowded, soft and grass-like ± 2–6 cm × 0.5–1.5 mm, with pale membranous sheathing bases and short apiculate tips ± 0.3–0.5 mm long, the narrow midrib evident on the underside; a few of the upper pairs of leaves much reduced with connate sheathing bases; *peduncles* 2–6 cm long, shorter than, subequal to, or usually overtopping the leaves in fruit; *flowers* solitary, terminal, 5-merous; *sepals* persistent, green with narrow

hyaline margins, ± 3–4 × 1–1.5 mm, about ⅔–¾ the length of the open capsule, ovate-triangular, obtuse to subacute (rarely mucronulate), often shortly incurved or blunt and ± thickened at the apex; *petals* 0; *stamens* 5, ± 1.5 mm long, alternate with the sepals; *anthers* 0.3–0.4 mm long; *capsule* ovoid, 5-valved, 3.5–4 × ± 3 mm; *seeds* smooth, obliquely subangular-pyriform in outline, pale reddish brown, ± 0.8–0.9 × 0.7 mm.

Alpine tracts of the Australian Alps, and mountains of Tasmania and New Zealand.

Distribution

Mainly in tall alpine herbfield and feldmarks; Moore (1970) has pointed out the close relationships between this species, *C. quitensis* (Kunth) Bartl. of South America, and *C. apetalus* (Labill.) Druce of south-eastern Australia, Tasmania and New Zealand. The last species is found occasionally in the subalpine tract as well as at lower altitudes, and differs from *C. affinis* mainly in the narrower stiff mucronate sepals which are subequal to or longer than the mature capsule and the stiffer leaves with longer acicular tips 0.5–1 mm long.

Notes and Habitat

Reference: Moore, D. M. (1970).

C. pulvinatus F. Muell., Trans. Phil. Soc. Vict. 1: 101 (1855)
'Hard cushion-plant'

Compact glabrous cushion-plant ± 2–10 cm diam.; *leaves* subulate, very stiff and ± spreading, the blade ± 2–4(–4.5) mm long excl. the acicular pungent tip (0.5–)1–1.5(–2) mm long, the margins and the midrib on the underside thickened and very conspicuous; *flowers* solitary, terminal, subsessile, the peduncles elongating to about 5 mm in fruit; *sepals* 5 (occasionally 4 or 6), erect, ± (1.5–)2–3.5 × 1–1.5 mm in the fruiting stage, subequal to or distinctly overtopping the opened capsule, ovate-triangular, rigid and tapering to a pungent tip, the midrib and upper marginal nerves thick and conspicuous; *petals* 0; *stamens* 5, alternate with the sepals, the filaments ± 1 mm long; *anthers* 0.3–0.4 mm long; *styles* 5, filiform, 0.5 mm long; *capsule* ovoid, 2–3 × ± 1.5 mm, 5 (occasionally 4 or 6)-valved, shorter than or subequal to the sepals; *seeds* red-brown, shining, ± 0.7–0.8 × 0.4–0.5 mm.

[162]

Endemic to the alpine tract of the Kosciusko area, N.S.W.

Distribution

This and the following species are the main representatives of the cushion-plant growth form on the Australian mainland. Such plants are particularly well developed in mountain environments in the southern hemisphere with cool moist 'oceanic' summers, including Tasmania, the South Island of New Zealand and South America. The rarity of the cushion-plant form on the Australian mainland is probably related to the occurrence of occasional warm dry spells in summer to which cushion-plants apparently are not adapted. Found mainly in *Epacris–Chionohebe* feldmark, *C. pulvinatus* has also extended its range into eroded tall alpine herbfields where it is an effective colonizer of bare stony ground. It appears to be closely related to some of the New Zealand species, e.g. *C. canaliculatus* Kirk, and also *C. hookeri* Cheesm. of Auckland and Campbell Islands.

Notes and Habitat

C. nivicola M. Gray, Contrib. Herb. Aust. No. 26: 8 (1976)
'Soft cushion-plant'

[163–4] Compact glabrous cushion-plant about 2–10(–15) cm diam.; *leaves* closely imbricate, suberect, the blade narrowly subulate, ± 2.5–5 × 0.5–0.8 mm, excluding the 0.4–0.7-mm acicular tip; *flowers* solitary, terminal, subsessile, the peduncles elongating to about 1 cm in fruit; *sepals* 5 (occasionally 4 or 6), ovate-triangular, acute, 1.5–3 × 0.8–1.3 mm in the fruiting stage, subequal to or shortly overtopping the opened capsule; *petals* 0; *stamens* 5, alternate with the sepals, the filaments ± 1 mm long; *anthers* 0.3–0.4 mm long; *styles* 5; *capsule* ovoid, ± 2 × 1.5 mm, 5 (occasionally 4 or 6)-valved; *seeds* red-brown, shining, ± 0.5–0.6 × 0.3–0.5 mm.

Distribution Endemic to the alpine tract of the Kosciusko area, N.S.W.

Notes and Habitat This species is sometimes difficult to distinguish from *C. pulvinatus* in the field where their habitats adjoin; however, its narrower, softer, suberect leaves tend to give the cushions a more moss-like appearance. It prefers conditions of greater snow cover and soil moisture, typically in *Coprosma–Colobanthus* feldmark and short alpine herbfield, and the cushions appear to be quicker-growing and show more tendency to coalesce, sometimes forming extensive mats or carpets. It also colonizes eroded areas in damp sites in tall alpine herbfields.

Stellaria L.

S. multiflora Hook., Hook. Comp. Bot. Mag. 1: 275 (1836)
'Rayless starwort'

[165] Small glabrous annual herb with weak decumbent or ascending stems, usually only a few centimetres high above the treeline (10 cm or more at lower elevations); *leaves* sessile or the lowest petiolate, opposite, ± 5(–10) mm long, narrowly elliptical to narrow lanceolate, acute; *flowers* solitary, axillary, on pedicels shorter or occasionally longer than the leaves; *sepals* 5, ± 4–5 mm long, ovate-lanceolate, shortly acuminate, usually 3-nerved, green with scarious margins; *petals* 0; *stamens* 5–10; *styles* 3, ± 3–5 mm long; *capsule* ovoid, 6-valved, subequal to or longer than the sepals at maturity; *seeds* reddish brown, ± 1 mm diam., densely covered with conspicuous blunt tubercles.

Distribution Widespread in temperate Australia, e.g. Western Australia, South Australia, Victoria, New South Wales, Australian Capital Territory and Tasmania.

Notes and Habitat Rare above the treeline on bare rocky areas. The beautifully sculptured tuberculate seeds are most attractive when viewed through a lens.

Ranunculaceae

Caltha L.

C. introloba F. Muell., Trans. Phil. Soc. Vict. 1: 98 (1855)
'Alpine marsh-marigold'

Glabrous tufted perennial herb ± 3–12 cm high, with copious long slightly fleshy   [166–72]
roots, the base of the plant surrounded by the persistent fleshy remains of old
leaf bases; *rhizomes* branched, erect or ascending, ± 0.5 cm diam.; *leaves* radical,
the blade ± 1–2.5 × 0.5–1 cm, narrowly ovate to ovate-lanceolate, obtuse or
emarginate, entire or very obscurely crenate, hastate or sagittate at the base
with two conspicuous narrow inflexed basal appendages ± 0.6–1.4 cm long,
or sometimes rounded or cordate at the base with the appendages attached only
at the junction of petiole and blade; *petioles* ± 2–6 cm long, the lower half with
very broad membranous sheathing margins which project upwards into a hood;
flowers very variable, sweetly scented, ± 3–6 cm diam., solitary on short fleshy
naked peduncles which elongate to 10 cm or more in fruit; *tepals* 5–8, usually
white or cream, sometimes variably tinged with magenta, narrow-lanceolate,
elliptic, or ovate-lanceolate, with a short ± twisted acuminate tip; *stamens* ±
12–25, the filaments flattened; *follicles* 5–15 or more, sessile, beaked, 1–5-seeded,
arranged in a whorl 1–2 cm diam.; *seeds* ± 1.5 × 1 mm, ovoid-ellipsoid, pale
brown, smooth and shining.

Alpine tract of the Australian Alps (see Notes below).   *Distribution*

Caltha is generally regarded as one of the most primitive genera of the Ranun-   *Notes and*
culaceae and is widespread throughout cold and cool temperate areas of both   *Habitat*
hemispheres. *C. introloba* is a characteristic species of short alpine herbfield, found
below snow patches and in shallow snow-melt streams. The flowers begin to
open even while still covered by snow. Smit (1973) maintains that *C. phylloptera*
A. W. Hill of Tasmania falls within the range of *C. introloba*.

References: Hill (1918); Smit (1973).

Ranunculus L.

KEY TO THE
SPECIES

1	Petals white or pale cream		2
1	Petals bright yellow (sometimes fading to white in old flowers)		3

2 Leaves orbicular in outline with numerous broad overlapping lobes; flowers 3–6 cm diam.; robust herb ± 15–30(–35) cm high   *R. anemoneus* (p. 174)
2 Leaves pinnately divided into narrow-linear segments; flowers 6–10(–17) mm diam.; dwarf herb 1–5(–8) cm high   *R. millanii* (p. 175)

3 Leaves entire, or with 1–2 shallow notches near the apex; blades with very coarse bristly hairs on the upper surface   4
3 Leaves deeply ternately or pinnately dissected   5

4 Marginal hairs of leaf blade appressed, blades usually shorter than the petioles; flowering stems exceeding the leaves   *R. muelleri* var. *muelleri* (p. 176)
4 Marginal hairs of the leaf blades ± spreading, the blades usually longer than the petioles; flowering stems usually shorter than the leaves (dwarf feldmark ecotype of the preceding)   *R. muelleri* var. *brevicaulis* (p. 176)

5 Nectary lobe (near base of petal) conspicuous   6
5 Nectary lobe absent   7

6 Petals (6–)10–14; leaf blades with linear-lanceolate lobes, hirsute with long patent hairs at least on the underside   *R. dissectifolius* (p. 177)
6 Petals usually 5 (rarely 6–8); leaf segments broad, covered with ± appressed hairs   *R. graniticola* (p. 177)

7 Leaves pinnately divided into numerous narrow-linear subterete segments; petals usually with 3(–5) nectaries   *R. gunnianus* (p. 178)
7 Leaves ternately or biternately dissected, the segments broad, flat and coarsely toothed or lobed; each petal with a single nectary   *R. niphophilus* (p. 178)

R. anemoneus F. Muell., Trans. Phil. Soc. Vict. 1: 97 (1855)
'Anemone buttercup'

[173–6]   Robust perennial herb to 35 cm or more high from a thick ascending or shortly creeping rootstock; *flowering stems* ± pilose or subglabrous, simple or branched above, usually with a few broad scales at the base and surrounded with the fibrous remains of old stems and petioles; *leaves* sparsely pilose above, pilose to silky-villous on the underside, or subglabrous, the basal leaves petiolate, the blades ± 5–8 cm diam., orbicular in outline, deeply palmately 3–5 cleft, the segments again variably digitately cleft into spreading and ± overlapping coarsely dentate lobes; *petioles* ± 5–15 cm long with broad sheathing bases; *cauline leaves* sessile and stem-clasping, the lowermost otherwise rather similar to the basal leaves, the upper becoming smaller with narrower segments; *flowers* white, solitary or 2–4, (3–)4–6 cm diam.; *sepals* 5–7, ± 1–1.5 cm long, ovate, acute, obtuse, apiculate or variably toothed or notched at the apex, silky-hairy on the back; *petals* usually numerous (± 20–30) and overlapping, narrowly obovate or obovate-cuneate, rounded to subacute at the apex, ± 1.5–3 cm long; *nectary* elongate, naked; *achenes* glabrous, numerous in a dense ovoid to subglobular head ± 1–1.5 cm diam., the body turgid, ± 3 mm long, the prominent ± 2-mm-long beak straight or recurved; *receptacle* sparsely hirsute, ± swollen at maturity.

Alpine and subalpine tracts of the Kosciusko area, N.S.W. (see Notes below.) *Distribution*

Locally common near snow patches both in short alpine herbfield and along *Notes and* snow-melt streams in tall alpine herbfield, also in rock crevices in *Copros-* *Habitat* *ma–Colobanthus* feldmark. Like *Caltha introloba*, *R. anemoneus* undergoes extensive development whilst still snow-covered, and is thus one of the first species to flower when the snow is melting. With its showy white flowers, it is one of the most spectacular buttercups in the world, and has strong affinities with some of the New Zealand alpine species, particularly with *R. buchananii* Hook. f. (Fisher 1965). Prior to the cessation of livestock grazing, *R. anemoneus* was almost grazed out of existence in the Kosciusko area, but it is now making a spectacular recovery there (Costin 1958). Stirling (1887) recorded this species from Mt Hotham in Victoria with the note: 'this plant is fast disappearing from the summits of our Victorian mountains, owing to the inroads made into the native vegetation by stock, as these alpine areas become increasingly occupied year by year'; however, there has been no subsequent confirmation of this record. The Monaro Conservation Society has adopted *R. anemoneus* as its symbol.

References: Costin (1958); Fisher (1965); Stirling (1887).

R. millanii F. Muell., Hook. Kew J. 7: 358 (1855)

'Dwarf buttercup'

Dwarf stoloniferous perennial herb 1–5(–8) cm high; *stolons* 1–3 cm long, some- [177–8] times becoming erect with leaves and flowers at the tip; *flowering stems* simple, usually sparsely covered with long spreading hairs; *leaves* radical, the laminae glabrous or almost so, pinnately divided into 3–5 narrow–linear segments 0.5–2 mm broad, each with a microscopic terminal pore, the lateral segments often bifid or trifid; *petioles* 1–8 cm long, sparsely hairy or glabrous, expanded and sheathing with membranous margins at the base; *flowers* solitary, 6–10(–17) mm diam., white or pale cream; *sepals* 5(–6), 2–4.5 mm long, sometimes hooded, pale green with hyaline margins, glabrous or with a few hairs on the back; *petals* usually 5 (rarely to 12), (3.5–)6(–8) × 2–4 mm, obovate or obovate-cuneate; *nectary* in a small semilunar pocket; *achenes* in a loose globular head 4–6 mm diam., elliptic, ovate or obovate in outline, sublenticular, 2–2.5 × 1.2–1.8 mm excl. the 0.5–1 mm straight or hooked beak; $2n = 16^{*}$.

Alpine and subalpine tracts of the Australian Alps and associated ranges as far *Distribution* north as Mt Gingera, A.C.T.

Usually found in small depressions subject to periodic inundation in sod tussock *Notes and* grassland and fen, and in wet mossy patches on the margins of bogs. *R. millanii* *Habitat* is believed to form natural hybrids with the following species: *R. graniticola*, *R. muelleri* var. *muelleri*, *R. dissectifolius*, *R. niphophilus*, and with other species at lower altitudes (Briggs 1962). Briggs (1959) indicates that this species sometimes tends to develop multiple nectaries.

References: Briggs (1959, 1962)*; Melville (1955).

R. muelleri Benth., Flor. Aust. 1: 13 (1863) var. *muelleri*
'Felted buttercup'

[179–80] Tufted perennial herb to 16 cm high; *flowering stems* simple, antrorsely strigose, usually overtopping the leaves; *leaves* mostly basal, petiolate, the blade ± 0.8–2.8 × 0.5–1.5 cm, elliptic or obovate to oblong, acute, entire or with 1–2 shallow notches near the apex (if ± deeply 3-lobed, see Notes below), densely antrorsely strigose on the margins and underside, the upper surface bristly with coarse, stiff, appressed or spreading tubercle-based hairs 2–4 mm long; *petioles* 1–9 cm long, grooved adaxially, densely antrorsely strigose, expanded and sheathing with membranous margins towards the base; *flowers* solitary, ± 1.5–2.5(–3.5) cm diam., bright golden yellow; *sepals* 5, ovate to elliptic, pale green and appressed-hairy on the back, usually glabrous on the membranous margins; *petals* usually 5, golden-yellow, obovate-cuneate, obtuse, truncate or emarginate; *nectary* with a small triangular lobe usually about 0.5 mm long, occasionally naked; *achenes* numerous in a dense subglobular head (7–)10(–14) mm diam., lenticular or ± swollen, the body ± 2–3 × 1.3–1.6 mm, oblong-elliptic to semi-orbicular in outline, the prominent beak 1–1.5 mm long, straight or slightly hooked at the apex; $2n = 16^\star$.

Distribution Alpine and subalpine tracts of the Kosciusko area, N.S.W., and the Bogong High Plains, Vic.

Notes and Common in tall alpine herbfields and sod tussock grassland. This species appears
Habitat to hybridize fairly freely with the following: *R. muelleri* var. *brevicaulis*, *R. graniticola*, *R. millanii*, *R. niphophilus*, *R. dissectifolius*, and with other species in Victoria (Briggs 1962). Specimens with ± deeply lobed leaves should be suspected of some degree of hybrid origin.

References: Briggs (1959, 1962)*.

R. muelleri Benth. var. *brevicaulis* B. G. Briggs, Proc. Linn. Soc. N.S.W. 84: 316 (1959)

[181–2] A dwarf feldmark ecotype of the preceding (var. *muelleri*) differing mainly as follows: *flowering stems* 2–4 cm long, subequal to or usually shorter than the leaves; *petioles* 0.5–2.5 cm long, subequal to or usually shorter than the blades, and the marginal hairs of the blades tend to spread, whereas they are appressed in the type variety; *achenes* semi-orbicular, with a more prominent beak 1.2–2.2 mm long; $2n = 16^\star$.

Distribution Known only from the summit feldmarks of the Kosciusko area, N.S.W., between Mt Kosciusko and Mt Twynam.

Notes and Fairly common in *Epacris–Chionohebe* feldmark. The two varieties are interfertile
Habitat and intermediate populations are found where their habitats adjoin (Briggs 1962).

References: Briggs (1959*, 1962).

R. dissectifolius F. Muell. ex Benth., Flor. Aust. 1: 11 (1863)

Perennial tufted herb 5–15(–25) cm high, usually forming dense clumps to 20 cm or more in diam.; *flowering stems* simple, spreading-hirsute at least proximally, usually antrorsely hirsute distally, subequal to or overtopping the leaves; *leaves* mostly basal, long-petiolate, the blade ovate to suborbicular in outline, densely hirsute with long spreading hairs on the underside, the upper surface glabrous or sparsely hairy, deeply ternately dissected, the segments again once or twice dissected into linear or linear-lanceolate lobes; *petioles* 2–12 cm long, densely hirsute with long spreading hairs, basally expanded with membranous margins and often purple-tinged; *flowers* solitary, bright golden yellow, 15–25(–32) mm diam.; *sepals* 5, 5–9 mm long, elliptic to obovate-cuneate, green and densely hirsute on the back, the margins membranous and usually glabrous; *petals* 6–10(–14), golden yellow, narrowly or broadly obovate-cuneate, truncate or emarginate at the apex; *nectary* with a small triangular or oblong lobe up to 1 mm long; *achenes* in a globular cluster ± 8–10 mm diam., the body 2–2.5 × 1.3–1.8 mm, lenticular or slightly swollen, gradually tapering to the 1–1.8 mm-long straight or curved beak; $2n = 16^\star$.

[184]

Alpine and subalpine tracts of the Kosciusko area, N.S.W.

Distribution

This species is mainly subalpine and is rather uncommon above the treeline. Briggs (1962) records natural hybrids between it and the following: *R. muelleri* var. *muelleri*, *R. graniticola*, *R. millanii* and with *R. pimpinellifolius* in the subalpine tract. It is mainly associated with bogs, but is also found in wet areas near streams in tall alpine herbfield and sod tussock grassland.

Notes and Habitat

References: Briggs (1959*, 1962).

R. graniticola Melville, Kew Bull. 1955: 205 (1955)

'Granite buttercup'

Perennial herb ± 5–20(–40) cm high, tufted or forming clumps to 20 cm or more in diam.; *flowering stems* usually simple, hirsute with spreading hairs or the hairs often antrorsely ascending or appressed above; *leaves* mostly basal, the blade with short appressed or ascending hairs, ovate or elliptic in outline, ternately or biternately dissected or lobed (see note on variation below), the ultimate segments spreading, lobed or dentate and obovate-cuneate in outline; *petioles* densely hirsute with long spreading hairs, expanded with membranous margins towards the base; *flowers* usually solitary, ± 15–25 (–30) mm diam.; *sepals* 5(–6), elliptic to ovate, green, hirsute on the back with glabrous membranous margins; *petals* 5(–8) golden yellow, broadly obovate-cuneate, obtuse, truncate or emarginate, sometimes tinged with purple on the back; *nectary lobe* 0.2–1 mm long, ± oblong, truncate or emarginate; *achenes* lenticular, obovate-cuneate or suborbicular, 2–3.5 mm long excluding the stout 0.7–1-mm-long curved beak, in a globular cluster 6–9 mm diam.; $2n = 16^\star$.

[185]

Alpine and subalpine tracts of the Australian Alps and adjacent ranges as far north as the Brindabella Range, A.C.T.

Distribution

Fairly common in tall alpine herbfields and sod tussock grassland. This species is very variable in the degree of dissection of the leaf in different parts of its

Notes and Habitat

range (Briggs 1959). Sometimes the flowers have almost black centres due to darkly stained styles and carpels, but this character is inconsistent even within a single population and many populations are entirely without dark-centred flowers. Briggs (1962) records natural hybrids between this species and the following: *R. millanii, R. dissectifolius, R. muelleri* var. *muelleri*, and with other species at lower altitudes.

References: Briggs (1959*, 1962).

R. gunnianus Hook., Hook. J. Bot. 1: 244, t. 133 (1834)
'Gunn's alpine buttercup'

[186]   Perennial herb 5–20(–25) cm high, the tough ascending or shortly creeping rootstock and the base of the plant densely sericeous with long antrorse hairs usually mixed with the fibrous remains of old petioles; *flowering stem* ebracteate or with a solitary linear bract, antrorsely pilose to almost glabrous, often sericeous towards the base and apex; *leaves* mostly basal, long-petiolate, the blades ± pilose or glabrescent, 2–3 pinnately divided into numerous linear subterete grooved segments which are usually tipped with a small pore; *petioles* grooved, ± 5–15 cm long, sheathing and ± sericeous at the base; *flowers* solitary, bright golden yellow, ± 2.5–4.5 cm diam.; *sepals* 5–6, subpetaloid, tinged and veined with purple on the back, shorter than the petals; *petals* 5–10(–13), shining golden yellow, fading to cream or white with age, tinged with pale purple on the outside, elliptical or obovate, usually with 3(1–5) naked nectary pits just above the short claw; *achenes* numerous in a dense subglobular cluster ± 1–1.5 cm diam., the body plump, obovoid-oblong, ± 2.5 × 1.5 mm, the prominent stout beak straight or slightly curved, ± 1–1.5 mm long.

Distribution   Alpine and subalpine tracts of the Australian Alps and mountains of Tasmania.

Notes and Habitat   Common in damp areas in alpine herbfields and sod tussock grassland. Like *R. anemoneus*, it has strong affinities with New Zealand alpine species, and according to Fisher (1965) bears some resemblance to *R. verticillatus* T. Kirk. It also seems to have affinities with the New Zealand *R. sericophyllus* Hook. f., a species of snowline fringe habitats, with which it shares the rather unusual character of multiple nectaries.

Reference: Fisher (1965).

R. niphophilus B. G. Briggs, Proc. Linn. Soc. N.S.W. 84: 321 (1959)

[187]   Perennial herb ± 5–15(–20) cm high, often forming dense bright green clumps to 30 cm or more in diam.; *flowering stems* simple, glabrous or hirsute; *leaves* mostly basal, glabrous or sparsely hirsute with long spreading hairs, the blade orbicular or deltoid in outline, ternate or biternate, the ultimate segments lanceolate, acute and usually lobed or dentate; *petioles* flattened and often purplish and with wide membranous margins towards the base; *flowers* solitary, ± 15–20(–30) mm diam.; *sepals* 5(–7), pale green, 5–8 × 2–4 mm, elliptic to ovate, glabrous or with sparse long hairs on the back; *petals* 5(–7), bright golden yellow,

7–14 × 4.5–9 mm, obovate-cuneate, truncate or emarginate, the nectary in a small pit with no distinct lobe; *achenes* lenticular or slightly swollen, ovate or elliptic in outline, ± 2.4–3 × 1.2–1.7 mm excluding the 1–1.7-mm-long straight or curved beak, numerous in a dense globular or conical cluster about 8–12 mm diam.; $2n = 16^*$.

Endemic to the Mt Kosciusko area, N.S.W., mainly in the alpine tract.   *Distribution*

Locally common and often forming extensive colonies in short alpine herbfield   *Notes and*
and in springs and flushes in tall alpine herbfields. Briggs (1962) records hybrids   *Habitat*
between this species and *R. millanii*, *R. muelleri* var. *muelleri*, and with *R. clivicola*
in the subalpine tract. We suggest that an appropriate common name for this
species would be Snow Buttercup.

References: Briggs (1959*, 1962).

Winteraceae

Tasmannia R. Br. ex DC.

T. xerophila (Parment.) M. Gray, Contrib. Herb. Aust. No. 26: 8 (1976)   'Alpine pepper'

Glabrous dioecious aromatic shrub about 0.5–1.5(–3) m high often forming ex-   [188–9]
tensive clumps from root suckers and rhizomes; *branchlets* reddish, with conspicuous leaf scars and rough with small dense rounded tubercles (colliculate); *leaves*
coriaceous, obscurely veined, shortly petiolate, narrowly oblanceolate to narrowly obovate, entire, up to 7(–10) cm long, subacute, obtuse or rounded at
the apex, the margins ± recurved; *petiole* densely tuberculate, 1–3(–5) mm long;
flowers pedicellate in terminal clusters, the sepals and petals minutely gland-
dotted, the 2 sepals fused in bud except at their apices, splitting along their margins
at anthesis; *male flowers: sepals* 2, broadly ovate to suborbicular (2.5–)5(–7.5)
mm long; *petals* 2, pale yellowish green, narrow elliptic to narrowly oblanceolate,
subequal to or slightly longer than the sepals; *stamens* usually numerous; a rudimentary sterile carpel usually present; *female flowers:* smaller than the male, the
stamens absent; *carpels* 1–5, shortly stipitate, the stigmatic crest oblique; *fruit*
of 1–3(–5) purplish black, shortly stipitate, ellipsoid to obovoid berries; *seeds*
black, shining, strongly curved, ± 2–2.5 × 1.5–2 mm; $n = 13$ (as *Drimys lanceolata*)*.

Alpine, subalpine and montane tracts of the Australian Alps and associated ranges   *Distribution*
as far north as the A.C.T.

Fairly common in heaths, mainly in moister sites. The leaves have a hot peppery   *Notes and*
taste if chewed, thus accounting for the common name. This species is extremely   *Habitat*
variable and the above description applies only to the Kosciusko alpine and subalpine populations.

References: Hotchkiss (1955)*; Vink (1970); Willis (1957).

Cruciferae

KEY   1   Fruit linear; leaves pinnate, the cauline leaves not amplexicaul
                                                        Cardamine (p. 180)
        1   Fruit triangular-obcordate; basal leaves toothed or pinnatifid, the cauline
            leaves sagittate-amplexicaul; locally common on Soil Conservation recla-
            mation areas and other disturbed sites
                                    Shepherd's Purse *Capsella bursa-pastoris*

Cardamine L.

[190–1]   The native species of *Cardamine* need revision (Thurling 1966 *a,b,c*, 1968). The
            common tufted form of herbfields and heaths above the treeline corresponds
            to species C of Thurling (1968), with a chromosome number of $2n = 48$. A
            robust form with large flowers and decumbent stems creeping and rooting at
            the base, which forms extensive colonies near late snow areas on the margins
            of streams and lakes, appears to be distinct.
            References: Thurling (1966 *a,b*,c; 1968).

Droseraceae

Drosera L.

D. arcturi Hook., Hook. J. Bot. 1: 247 (1834)   'Alpine sundew'

[192]   Tufted insectivorous perennial herb ± 2–6(–10) cm high, the base of the plant
            surrounded by the remains of old leaves; *rootstock* erect or ascending, sometimes
            branched and creeping through mosses; *leaves* radical, ± 2–7 cm × 3–5 mm,
            the blade narrowly oblong to linear, obtuse or rounded at the apex, the upper
            surface covered with long glistening viscid gland-tipped hairs (tentacles) and
            minute sessile glands, slightly narrowed to a glabrous pseudo-petiole which is
            almost as long as the blade and sheathing at the base; *scapes* slender, solitary
            or few, usually over-topping the leaves, sometimes with a small bract in the
            upper part; *flower* solitary, white, ± 1 cm diam.; *sepals* 5, persistent, ± 5–7 ×
            2 mm, shortly united at the base, narrow elliptic or oblong, obtuse to subacute,
            sometimes notched or erose at the apex, green becoming blackish when dried;
            petals 5, free, white, 6–10 mm long, broadly ovate or obovate, rounded at the

apex; *stamens* 5, shorter than the petals and alternating with them; *anthers* extrorse; *ovary* superior, unilocular; *styles* 3–4, ± 0.5 mm long with capitate stigmas; *capsule* ellipsoid, 0.6–1 cm long, loculicidal.

Alpine and subalpine tracts of the Australian Alps, and mountains of Tasmania and New Zealand (North and South Islands and Stewart Island). *Distribution*

Fairly common in wet areas, as in bogs and short alpine herbfield. The sticky hairs on the leaves trap small insects which are digested by powerful leaf exudates and absorbed into the plant as a source of nitrogen. *Notes and Habitat*

Reference: Erickson (1968).

Crassulaceae

Crassula L.

C. sieberana (Schult. & Schult. f.) Druce, Rep. Bot. (Soc.) Exch. Cl. Manchr 1916: 618 (1917)

Small succulent annual herb, green or reddish-tinged, up to a few cm high (to 12 cm at lower altitudes); *leaves* crowded, ± 2–3(–4) mm long, opposite and connate at the base, thick and fleshy, ovate-oblong in outline, blunt; *flowers* 4-merous, ± 2 mm diam., subspicate in ± dense axillary clusters or occasionally solitary in the axils, subsessile or sometimes 1 flower pedicellate; *calyx* divided almost to the base into ovate-triangular acute or acuminate lobes 1–1.5 mm long; *petals* hyaline, pinkish, ± 1 mm long, lanceolate, acuminate; *stamens* 0.5 mm long, alternate with the petals; *nectary scales* linear-spathulate, ± 0.4 mm long; *fruiting carpels* splitting open along the inner face, ± 1 mm long incl. the short beak formed by the persistent style; *seeds* usually 2 in each carpel, narrowly ellipsoid, ± 0.5 mm long. [193]

Temperate eastern Australia, Tasmania and New Zealand. *Distribution*

Rare above the treeline, occasionally found colonizing bare areas in feldmark or rocky cliffs overlooking lakes etc. Healy (1948) has pointed out that although this species is usually described as lacking nectary scales, all the New Zealand specimens referred to this species which were examined by him were found to possess nectary scales. This appears to be also the case with regard to Australian specimens, the scales being usually stuck firmly to the inside of the petals in dried specimens, and thus easily overlooked. *Notes and Habitat*

Reference: Healy (1948).

Rosaceae

KEY   1   Flowers sessile, crowded in dense globular heads; hypanthium with 4 con-
            spicuous barb-tipped spines; leaves pinnate   *Acaena* sp. (p. 182)
        1   Not as above   2

        2   Robust perennial herb ± 15–20 cm high from a thick woody stock; flowers
            in loose terminal cymose clusters; leaves reniform, palmately lobed with 9–11
            shallow rounded lobes; stamens 4   *Alchemilla xanthochlora* (p. 183)
        2   Small annual herb; flowers in small condensed leaf-opposed clusters; leaves
            fan-shaped, deeply dissected; stamen 1; casual on Soil Conservation recla-
            mation areas, probably not persisting   Parsley Piert*Aphanes arvensis

Acaena Mutis ex L.

A. sp.

[194]   Prostrate creeping much-branched undershrub usually forming dense dark green
        patches up to 1 m or more in diam.; *main stems* creeping and rooting at the
        nodes, sparsely appressed-hairy, soon glabrescent; *flowering stems* erect or ascen-
        ding to ± 15–20 cm high in fruit; *leaves* pinnate with 3–5 pairs of leaflets, ±
        pilose on the petiole and rachis; *leaflets* sessile or the lower shortly stalked, the
        two distal pairs elliptic or obovate or broadly elliptic or broadly obovate to subor-
        bicular in outline, rounded or truncate at the apex, asymmetrical at the base,
        the margins coarsely crenate-serrate with 10–14 teeth, shiny dark green and
        glabrous above, paler on the underside with inconspicuous appressed hairs on
        the midrib and usually on the veins; *stipules* adnate, with free entire or toothed
        herbaceous apices; *flowers* sessile, crowded in dense solitary globular heads on
        terminal sparsely hairy or ± glabrous peduncles which elongate to 5–10 cm
        in fruit; *sepals* 4, herbaceous, ovate to elliptic, ± 1.5 mm long, sparsely hairy
        on the back, persistent in fruit; *petals* 0; *stamens* 2, exserted; *anthers* ± 0.5 mm
        long; *style* short, the stigma feathery; *fruiting heads* 1.5–2.5 cm diam., incl. the
        spines; *fruiting hypanthium* pilose, obconic, 2.5–3 mm long, 4-angled, each angle
        prolonged into a shining reddish purple barb-tipped spine 4.5–8 mm long; $2n$
        = 42 (as *A. anserinifolia*)*.

Distribution   See Notes below.

Notes and   Fairly common in heaths and in rocky situations in sod tussock grassland and
Habitat   tall alpine herbfields, this plant is easy to recognize by its red globular inflor-
                 escences and the propensity of the numerous spiny seed-cases to adhere to socks
                 and other clothing. Yeo (1973) has indicated that the complex in Australia which
                 has been referred to by most authors as *A. anserinifolia* is in need of revision.
                 The above description applies to the high-altitude form of this complex which
                 is generally much less hairy and with broader leaflets than forms from surround-
                 ing lower elevations.

References: Moore (1964)*; Yeo (1973).

Alchemilla L.

A. xanthochlora Rothm., Feddes Repert. 42: 167 (1937)
'Lady's mantle'

Perennial herb ± 15–20 cm high from a thick woody ascending or shortly creep- [195]
ing rootstock, the base of the plant surrounded by the remains of old leaves;
stems densely clothed with ± spreading hairs, the upper parts of the inflorescence
glabrous or almost so; *leaves* mostly radical, long-petiolate, reniform, palmately
lobed with 9–11 shallow rounded lobes, the margins serrate with curved subequal
red-tipped teeth, the upper leaf surface glabrous or with lines of sparse appressed
hairs along the folds to the lobe sinuses, hairy on the veins and sometimes sparsely
on the surface beneath; *petioles* with dense ± spreading hairs and brown scarious
adnate stipules; *flowers* small, greenish, 4-merous, ± 2.5 mm diam., in loose
cymose clusters; *hypanthium* urceolate, glabrous or almost so; *sepals* ovate-
triangular, ± 1 mm long, the epicalyx segments about half as long; *petals* 0;
stamens usually 4, alternating with the sepals; *style* basally attached on the single
carpel; *achene* brown, ovoid, ± 1.5 × 1 mm, enclosed in the hypanthium.

Alpine and subalpine tracts of the Australian Alps; widespread in Europe and *Distribution*
naturalized in eastern North America.

Occasionally found above treeline in small colonies in moist rocky sites in tall *Notes and*
alpine herbfield, as on the crags at the head of Blue Lake and Club Lake and *Habitat*
along the north bank of Lake Albina. This species may have been a very early
introduction, but the evidence of its ecological distribution in remote, undisturbed and specialized habitats and its early presence in the Australian Alps is
very puzzling. Ferdinand Mueller, in a letter to W. J. Hooker from Omeo dated
16 December, 1854, wrote: 'You may imagine, Sir William, what a hearty wel-
come our old acquaintance *Alchemilla vulgaris* had when I found a few individuals
of it here in the very heart of the Alps, viz. at the sources of the Mitta Mitta,
not having seen this plant during the last seven years, when I left my native
home.'

References: Mueller (1855*a*); Rothmaler (1955).

Papilionaceae

| 1 | Leaves simple; shrubs or subshrubs | 2 | KEY |
| 1 | Leaves trifoliolate; herbs | 3 | |

| 2 | Flowers deep violet-blue or white; stamens united by their filaments into a tube open on the upper side | *Hovea purpurea* var. *montana* (p. 184) |
| 2 | Flowers orange-yellow, variably tinged with red; stamens free | *Oxylobium* (p. 185) |

3 Heads usually sessile within an involucre of the uppermost stipules; corolla deep pinkish red to reddish purple; robust ± villous perennial with large globose to ovoid heads ± 2–2.5 cm wide; rare casual, roadsides and waste places   Red Clover * *Trifolium pratense*
3 Heads pedunculate, not involucrate; corolla white or pinkish   4

4 Glabrous creeping perennial rooting at the nodes; leaflets obovate or obcordate; corolla exceeding the calyx; heads globular, ± 15–20 mm wide; locally common on disturbed sites and Soil Conservation reclamation areas
White Clover *Trifolium repens*
4 Softly hairy ± erect slender annual; leaflets narrowly obovate-oblong, obtuse or emarginate; corolla much shorter than the calyx; heads softly downy, at length ± cylindrical; rare casual, roadsides and disturbed areas
Hare's-foot Clover *Trifolium arvense*

Hovea R. Br. ex Ait.

H. purpurea Sweet var. *montana* Hook. f., Flor. Tas. 1: 93 (1856)
'Alpine hovea'

[196]   Low spreading shrub ± 15–30(–45) cm high, sometimes forming patches up to 1 m or more in diam.; *stems* tough and wiry, decumbent or ascending, rooting at the base, the branchlets shortly tomentose, glabrescent, the young shoots ferruginous; *leaves* alternate, shortly petiolate, coriaceous, 10–25 × 3–6 mm, oblong-elliptic, obtuse, sometimes with a small blunt apiculum, the upper surface glabrous, glossy green, obscurely reticulate, the underside with thin silver-grey or pale ferruginous tomentum, the midrib impressed above and prominent on the underside, the margins recurved or revolute; *petioles* tomentose, 1.5–3.5 mm long; *stipules* ± 1 mm long; *flowers* deep violet-blue or white, solitary or paired in the upper axils on short peduncles ± 2 mm long; *bract* and *bracteoles* blunt, ± 2 mm long; *calyx* 4–4.5 mm long, ferruginous-tomentose, 2-lipped, the upper lip broad, subtruncate, deeply notched, the lower lip with 3 narrowly triangular lobes; *standard* with suborbicular emarginate lamina ± 0.8 mm wide, the wings and keel shorter, ± 1 cm long; *stamens* 10, the filaments all united in a tube open along the upper side; *ovary* pubescent; *ovules* 2; *pod* swollen, ferruginous-tomentose, ± 1 × 0.8 cm, obliquely oval to suborbicular in outline, shortly pilose inside, the style base persistent.

Distribution   Alpine and subalpine tracts of the Australian Alps and associated ranges as far north as the A.C.T., and mountains of Tasmania.

Notes and Habitat   This species is an important component of subalpine heaths, but is marginal above the treeline.

Oxylobium Andr.

O. ellipticum (Labill.) R. Br. ex Ait.f., Hort. Kew., ed. 2, 3: 10 (1811)   'Common oxylobium; golden shaggy-pea'

Variable shrub, procumbent and rooting near the base, or erect and up to 2 m or more high at lower altitudes; *branchlets* tough, wiry, pubescent; *leaves* shortly petiolate, exstipulate, coriaceous, mostly in irregular whorls of 3 or 4, ± 1–2(–3) × 0.3–0.7 cm, elliptical, ovate or lanceolate with strongly recurved margins, mucronate, the upper surface appressed-hairy but soon glabrescent, dark green, strongly reticulate-veined and rough with minute tubercles which extend to the areas between the veins, the underside silvery appressed-pubescent; *flowers* orange-yellow, usually with a reddish blotch at the base of the standard and variably reddish-tinged on the keel, in short dense terminal or subterminal corymbs, subtended by small deciduous bracts; *bracteoles* deciduous, on the pedicels or near the base of the calyx; *calyx* appressed silky-pubescent, the lobes acuminate, subequal to or longer than the tube; *pod* shortly stipitate, villous, pale yellowish brown or greyish brown, 7–10 × 4–6 mm including the abruptly acuminate apex, ovoid, swollen, sometimes finely transversely wrinkled, glabrous inside.

Tablelands and mountains of eastern N.S.W., Victoria and Tasmania.

A characteristic component of subalpine heaths, but not ascending far above treeline. This is a rather variable species, the procumbent form from high altitudes has been separated as var. *alpinum* Maiden & Betche in Proc. Linn. Soc. N.S.W. 23: 11 (1898).

Reference: Thompson (1961).

[197–8]

Distribution

Notes and Habitat

O. alpestre F. Muell., Trans. Phil. Soc. Vict. 1: 38 (1855)   'Alpine oxylobium; mountain shaggy-pea'

Erect to decumbent shrub with tough wiry branches, the branchlets appressed-pubescent, soon glabrescent; *leaves* shortly petiolate, coriaceous, opposite or in whorls of three, 1–4 × 0.3–1.3 cm, oblong elliptic or lanceolate, mucronate, reticulate-veined, the upper surface dark green, appressed-hairy but soon glabrescent with minute tubercles restricted almost wholly to the veins, paler, sparsely hairy and glabrescent on the underside, the margins recurved; *stipules* subulate, 1.5–3 mm long, ± recurved; *flowers* ± 1–1.2 cm long, orange-yellow, in loose corymbose clusters; *calyx* appressed silky-pubescent, subtended by two linear

[199]

bracteoles about as long as the tube, the acuminate lobes about as long as the tube; *standard* orange-yellow with a reddish blotch at the base, the wings and keel variably red-tinged; *pod* shortly stipitate, hirsute, dark grey, 1–1.5 cm long, ovoid-ellipsoid, acuminate, the valves coarsely transversely ridged, lined on the inside with pale crinkly scaly hairs when freshly opened.

Distribution   Mountains of eastern Victoria and the Southern Tablelands of N.S.W., as far north as the A.C.T.

Notes and Habitat   This species is very marginal in the alpine tract, but quite common in the sub-alpine and montane tracts.

Reference: Thompson (1961).

Geraniaceae

Geranium L.

KEY TO THE SPECIES

1   Flowering stems short and rudimentary or reduced to a single peduncle and pedicel; pedicels with coarse dense antrorse-appressed and ± spreading hairs   *G. antrorsum* (p. 186)

1   Flowering stems well developed, leafy; pedicels pubescent with retrorse hairs   2

2   Seeds with shallow inconspicuous alveoli; bracteoles inserted about the middle of the pedicel-peduncle   *G. potentilloides* var. *potentilloides* (p. 187)

2   Close to the preceding, but seeds with larger, more conspicuous alveoli; bracteoles inserted in the lower half or near the base of the pedicel-peduncle
G. potentilloides var. *abditum* (p. 188)

G. antrorsum Carolin, Proc. Linn. Soc. N.S.W. 89: 357 (1965)

'Rosetted crane's-bill'

[200]   Tufted perennial herb to 15 cm high with thick simple or branched tap-root; *rootstock* short, branched, densely covered with the persistent brown remains of stipules and petioles; *flowering stems* very short and inconspicuous or reduced to a pedicel and peduncle; *leaves* mostly basal, petiolate, crowded, the blade semi-orbicular to cuneate in outline, 1–4 × 1.5–4.5 cm, with coarse appressed or subappressed hairs on both surfaces, palmately 5–7-lobed or dissected with each lobe again 3-lobed at the apex; *petioles* 3–16 cm long, densely covered with antrorsely appressed and ± spreading hairs; *stipules* 4–8 mm long, acuminate; *pedicel–peduncle* shorter than the leaves, 1–4 cm in flower and up to 10 cm in fruit, densely covered with coarse antrorse appressed hairs and variable ± spread-

ing hairs; *bracteoles* ± 6 mm long; *flowers* solitary; *sepals* 5, 5–11 × 2–4 mm incl. the conspicuous 1–2-mm-long awn, appressed-pubescent with coarse spreading hairs on the veins and margins; *petals* 5, deep pink, 6–12 mm long, obovate; *stamens* 10; *filaments* 2–3.5 mm long, expanded and ciliate at the base; *anthers* ± 1 mm long; *mericarps* brown, covered with stiff spreading hairs; *rostrum* ± 11 mm long; *seeds* dark brown or black, 2.5–3 × ± 1.5 mm, oval in outline, obscurely reticulate; *raphe* lateral.

Alpine, subalpine and cold valleys of the montane tracts of the Australian Alps and associated ranges as far north as the A.C.T. *Distribution*

Fairly common in the subalpine tract in tall alpine herbfields, sod tussock grass-land and on the margins of bogs, but less common above the treeline. *Notes and Habitat*

Reference: Carolin (1967).

G. potentilloides L'Hér. ex DC., Prodr. 1: 639 (1824)
var. *potentilloides*

Perennial herb ± 10–25 cm high with slightly thickened simple or branched tap-roots; *rootstock* simple or with few branches, covered distally with the persist-ent brown remains of stipules and petioles; *flowering stems* well developed, leafy, decumbent or ascending, pubescent with short retrorse or spreading hairs or glabrescent, sometimes rooting at the lower nodes; *basal leaves* similar to the cauline leaves but larger, usually not persisting; *cauline leaves* opposite, petiolate, the blade semi-orbicular to broadly ovate in outline, ± cordate at the base, ± 1–3 × 1–5 cm, hairy on both sides, paler (or sometimes purplish) on the underside, deeply palmately 5–7-lobed, the lobes oblong or obovate-cuneate in outline and usually irregularly 3-lobed at the apex; *petioles* 1.5–3.5 cm long, pubescent with retrorse and spreading hairs; *stipules* 3–10 mm long, acuminate, sometimes 2-fid at the apex; *pedicel–peduncle* 2–5 cm long, pubescent with soft retrorse hairs, geniculate at the bracteoles at maturity; *bracteoles* 2.5–4 mm long, inserted about the middle of the pedicel-peduncle; *flowers* solitary (rarely in pairs); *sepals* 5, 4–6× 1.5–2.5 mm incl. the short 0.5–1-mm-long awn, pubescent with short antrorsely appressed hairs and some longer spreading ones; *petals* 5, obovate, 5–6 × ± 3 mm, pink or white, ciliate at the base; *stamens* 10; *filaments* ± 3 mm long, expanded and ciliate at the base; *anthers* ± 0.5 mm long; *mericarps* brown, covered with stiff spreading hairs; *rostrum* ± 9 mm long; *seeds* dark brown, ± 2 mm long, with shallow elongate alveoli; *raphe* lateral. [201]

Widespread in eastern N.S.W., Victoria, South Australia and Tasmania; also New Zealand, subantarctic islands and east New Guinea. *Distribution*

A rather common, if inconspicuous, species of rocky sites in tall alpine herbfields and sod tussock grassland. *Notes and Habitat*

References: Carolin (1965, 1967).

G. potentilloides L'Hér. ex DC. var. *abditum* Carolin, Proc. Linn. Soc. N.S.W. 89: 340 (1965)

Close to var. *potentilloides* but usually smaller above the treeline, differing mainly in the larger more conspicuous alveoli of the ripe seeds, which are usually black when fully ripe, the shorter peduncles due to the bracteoles being inserted in the lower half or near the base of the pedicel-peduncle, and the often closer indumentum which is almost canescent in the upper parts of the plant.

Distribution   Alpine and subalpine tracts of the Australian Alps.

Notes and   Less common above the treeline than the var. *potentilloides*, and usually found
Habitat   in open areas in heaths. Carolin (1965) suggests that *G. potentilloides* var. *abditum* may hybridize with *G. sessiliflorum* Cav. subsp. *brevicaule* (Hook.) Carolin where their habitats adjoin in the subalpine tract.

Reference: Carolin (1967).

Rutaceae

Phebalium Vent.

P. ovatifolium F. Muell., Trans. Phil. Soc. Vict. 1: 99 (1855)

'Ovate phebalium'

[202–3]   Aromatic shrub 0.5–1 m high with tough spreading branches, sometimes growing close over rocks and boulders; *branchlets* rough with conspicuous raised oil-glands and densely covered with silvery grey or pale ferruginous scurfy scales; *leaves* flat, entire, alternate, shortly petiolate, coriaceous, broadly ovate or oval, ± 10(–15) × 7–8 mm, obtuse to rounded at the apex, the upper surface glabrous, glossy, olive-green and usually dotted with oil glands, densely scaly on the underside; *flowers* 1–3 in the upper axils, ± 1 cm diam., often pink-tinged in bud, on flattened scaly bracteate peduncles much shorter than the leaves; *sepals* 5, free, gland-dotted, triangular, ± 1.5 mm long, glabrous or one scaly, closely subtended by 2 scaly bracteoles; *petals* 5, white inside, imbricate, 4–5 mm long, elliptical, glabrous, with sparse oil-dots on the back and slightly inflexed tips; *stamens* 10, slightly shorter than the petals, the filaments glabrous; *anthers* pale purplish, 0.5–0.8 mm long; *ovary* broadly conical, densely scaly, subtended by a dark mauve disc 0.5 mm high; *style* simple, ± 2 mm long; *fruit* of 5 shortly beaked cocci ± 3 mm high, opening along the upper margin; *seeds* ellipsoid or subreniform, ± 2 × 1.5 mm.

Distribution   Restricted to the alpine and subalpine tracts of the Kosciusko area, N.S.W.

Notes and   One of the commonest shrubs above treeline and a major component of heaths.
Habitat   When growing near rocks, it often assumes espalier and rock-clinging habits; in more open sites, the branches are spreading and crowns tend to be flat-topped.

Reference: Wilson (1970).

Stackhousiaceae

Stackhousia Sm.

S. pulvinaris F. Muell., Trans. Phil. Soc. Vict. 1: 101 (1855)
'Alpine stackhousia'

Glabrous much-branched perennial forming dense dark green cushions or mats   [204]
to 5 cm high; *stems* slender, close-set, creeping and rooting, rough with conspicuous leaf scars towards the base; *leaves* crowded, bright green, thick and rather
fleshy, ± 5–10 × 1–2 mm, narrow-oblong or narrow-obovate, obtuse; *stipules*
minute, ± 0.5 mm long, usually persistent on either side of the leaf scars; *flowers*
5-merous, 8–10 mm diam., creamy yellow, sweetly scented, solitary and almost
sessile in the upper axils and only shortly overtopping the leaves; *pedicels* very
short with 1 or 2 minute bracts; *calyx* 5-lobed, ± 2 mm long, the lobes about
as long as the tube; *petals* 5, connate for the most part into a long tube ± 5
mm long, but free at the base; *stamens* 5, free, alternate with the petals; *anthers*
± 1 mm long, three extending to near the summit of the tube on long filaments,
two on shorter filaments extending about half-way up the tube; *nutlets* usually
3, ± 2–3 × 1.5 mm, yellowish becoming tinged with red when ripe, coarsely
wrinkled when dry.

Alpine and subalpine tracts of the Australian Alps, and local in montane and   *Distribution*
subalpine areas of Tasmania.

This mat plant is fairly common in moist sites in sod tussock grassland and tall   *Notes and*
alpine herbfields. Ferdinand Mueller, when he first described this plant in 1855,   *Habitat*
wrote: 'On the highest summits of the Australian Alps, where saturated with
moisture, the widely expanded tufts decorated with fragrant starry flowers form
a beautiful carpet'.

Violaceae

1	Herb; flowers irregular, 1–1.5 cm diam.; fruit a capsule	KEY
	Viola betonicifolia (p. 190)	
1	Depressed shrub; flowers regular, 3–4 mm diam.; fruit a berry	
	Hymenanthera dentata var. *angustifolia* (p. 190)	

Viola L.

V. betonicifolia Sm., Rees Cyclop. 37: 1 (1817)
subsp. *betonicifolia*   'Showy violet'

[205] Tufted acaulescent perennial herb 3–10(–15) cm or more high from a short erect or ascending rootstock; *leaves* radical glabrous, petiolate, the blades ± 1.5–5(–8) × 0.5–2.5 cm, triangular-ovate or oblong to lanceolate, obtuse or rounded at the apex, long-decurrent on the petiole, the margins shallowly and distantly crenate; *petiole* subequal to or up to twice as long as the blade; *stipules* adnate to the base of the petiole, the free points acuminate; *peduncles* subequal to or longer than the leaves, with two linear-subulate bracteoles attached below the middle; *flowers* solitary, nodding, irregular, ± 1–1.5 cm diam., pale purplish to almost white or deep violet with darker veins; *sepals* ovate to ovate-lanceolate, 3–6.5 mm long, green with scarious margins, imbricate, with short blunt basal appendages; *petals* 2–3 times as long as the sepals, obovate-oblong, shortly clawed, the lateral ones bearded inside, the lower pouched at the base; *anthers* 5, connivent around the ovary, the connectives expanded into scarious apical appendages, the two lower anthers spurred on the back; *style* ± 1.5(–2.5) mm long, geniculate at the base, clavate distally; *capsule* loculicidal, 7–10 mm long, ellipsoid or oblong, opening to the base in 3 boat-shaped valves; *seeds* ovoid, ± 1.5 mm diam.; $2n = 48\star$.

Distribution   Eastern Australia from Queensland to Tasmania, and rare in the Mt Lofty Ranges, South Australia; also widespread in south-east Asia and Malesia.

Notes and   A widespread native violet, common in tall alpine herbfields and sod tussock
Habitat   grassland.

References: Jacobs and Moore (1971); Moore (1963)$\star$.

Hymenanthera R. Br.

H. dentata R. Br. ex DC. var. *angustifolia* (R. Br. ex DC.) Benth., Flor. Aust. 1: 105 (1863)

[206] Depressed divaricate spiny shrub ± 15–50 cm high in rocky crevices or scrambling over rocks; *stems* prostrate or decumbent, rooting at the base; *bark* lenticellate; *branchlets* puberulent, glabrescent usually ending in a stout rigid spine; *leaves* glabrous, alternate or fasciculate, ± 1–3 cm long, subsessile or shortly petiolate, narrowly obovate-cuneate to oblanceolate, obtuse or rounded at the apex, the margins slightly recurved, distantly toothed or subentire, dark green above, paler on the underside; *stipules* minute, fimbriolate, caducous; *flowers* usually 5-merous (sometimes unisexual?), pale yellowish, usually solitary, ± 3–4 mm diam., nodding on short recurved pedicels which are about 1 mm long with 2 small bracteoles near the middle; *sepals* orbicular, 1–2 mm long, imbricate, ciliolate, persistent; *petals* imbricate at the base, about twice as long as the sepals, recurved above

the middle; *anthers* sessile, ± 1 mm long, cohering in a tube around the ovary, each with a cuneate dorsal scale, the connectives produced above into fringed scarious appendages about as long as the anther; *style* persistent, ± 0.5 mm long, with a flattened irregularly lobed capitate stigma; *berry* subglobular, slightly flattened, dark purple when ripe, 5–7 mm diam.

Tablelands and ranges of eastern N.S.W., Victoria, south-eastern South Australia, and Tasmania.

Distribution

The above description applies only to the alpine and subalpine form of this very polymorphic and widespread species. The typical habitat is at the base of rocks as a component of heaths. Green (1970) has followed Beuzenberg (1961) and united *Hymenanthera* with *Melicytus*.

Notes and Habitat

References: Beuzenberg (1961); Green (1970).

Thymelaeaceae

Drapetes Banks ex Lam.

D. *tasmanicus* Hook. f., Hook. Kew J. 5: 299, t. 7 (1853)

Small creeping perennial to about 5 cm high forming tufts or patches up to about 30 cm diam.; *stems* slightly fleshy, creeping and rooting, 2.3 mm diam., often much-branched and intertwined; *branchlets* lax, ascending, sometimes reddish, with prominent leaf scars towards the base; *leaves* stiff, linear, sessile, 1.5–3 × 0.2–0.6 mm, suberect and imbricate, blunt, with the margins ciliate above and a conspicuous tuft of hairs at the apex, glabrescent with age; *flowers* white, sessile, ± 3 mm diam., pilose outside, 1–4 in small terminal clusters among the upper leaves; *receptacle* villous; *floral tube* ± 1.5 mm long; *calyx lobes* 4, ovate, obtuse, subequal in length to the floral tube, each with 2 globular, sessile, yellow gland-like petaloid appendages inside at the base; *stamens* 4, ± 1 mm long, shortly exserted, attached at the top of the tube, alternate with the calyx lobes; *anthers* basifixed, 0.3–0.4 mm long; *ovary* glabrous or with a few long hairs towards the summit; *style* lateral, ± 1 mm long; *stigma* capitate, shortly exserted; *fruit* a small ovoid or ellipsoid drupe 1.7–2 × 1–1.2 mm.

[207–8]

Alpine tract of the Kosciusko area, N.S.W., and high mountains of Tasmania.

Distribution

Fairly common as a characteristic species of *Epacris–Chionohebe* feldmark.

Notes etc.

Pimelea Banks & Soland. ex Gaertn.

P. axiflora F. Muell. ex Meissn. var. *alpina* F. Muell. ex Benth., Flor. Aust. 6: 26 (1873)

[209] Small erect or ascending subshrub 15–40 cm high, glabrous except the inflorescence; *leaves* opposite, shortly petiolate, subcoriaceous, ± 8–15 × 3–7 mm, narrowly to broadly ovate to elliptic, obtuse or rounded at the apex, the midrib prominent on the underside; *petioles* 0.5–1 mm long, articulate, leaving a conspicuous scar when the leaves have fallen; *flowers* creamy white, appressed-pubescent on the outside, hermaphrodite or female, apparently functionally dioecious or almost so, in small sessile axillary clusters shorter than the leaves; *involucral bracts* 2–4, thin, much smaller than the leaves, subtended by a few small brown bractlets; *receptacle* shortly hairy; *hermaphrodite flowers*: *floral tube* 4–6 mm long; *calyx lobes* 2–3 mm long, broadly ovate or broadly oblong-elliptic; *stamens* 2, inserted near the throat, shortly exserted on filaments ± 0.5 mm long; *anthers* 0.6–0.9 mm long; *style* included, 1.5–2 mm long, the stigma small; *female flowers*: *floral tube* 3–4 mm long, tardily circumciss above the ovary, the lobes 1.5–2 mm long; *stamens* small and sterile; *style* exserted, the stigma prominent; *fruit* ovoid, 2.5–3.5 × ± 1.5 mm, enclosed in the persistent membranous base of the floral tube.

Distribution — Alpine and subalpine tracts of the Australian Alps.

Notes and Habitat — Occasional in heaths, tall alpine herbfield and sod tussock grassland, but not common above the treeline. As with *P. alpina*, the breeding system needs further study.

P. ligustrina Labill., Nov. Holl. Plant. Specim. 1: 9, t. 3 (1804) sens. lat.

[210] Shrub to about 1 m high, glabrous except the inflorescence; *leaves* flat, opposite, subsessile, narrowly or broadly elliptic to ovate, ± 1.5–3.5(–5) × 0.5–2 cm, often increasing in size towards the inflorescence, penniveined, paler on the underside; *inflorescence* of numerous flowers in globular terminal heads 2.5–4 cm diam., on short stalks 0.5–3 cm long; *involucral bracts* (6–)8, glabrous outside, appressed-pubescent inside, the margins fringed with long acicular hairs, the

larger outer bracts broadly ovate to suborbicular, ± 1.5–2 × 1–1.2 cm, usually strongly tinged with magenta, the inner bracts smaller, ovate, ± 1.2–1.5 × 0.6–0.7 cm; *receptacle* of the inflorescence densely covered with acicular hairs; *flowers* creamy white, the floral tube 10–12 mm long, circumciss above the ovary, appressed silky-villous, the lower part surrounding the ovary also covered with ± deciduous acicular hairs; *calyx lobes* ± 4 mm long; *stamens* long-exserted, the filaments ± 5 mm long; *anthers* 1.3–1.5 mm long, orange drying yellow; *style* slender, exserted subequal to or overtopping the anthers; *fruit* ovoid or ellipsoid, acuminate, hairy at the apex, 3.5–5.5 × 1.5–2.5 mm, enclosed in the persistent base of the floral tube.

See Notes below.

Distribution *Notes and* *Habitat*

The above description applies to the short alpine and subalpine form of this species which is found in the Australian Alps in N.S.W. and Victoria. The species in the broad sense is found from south-eastern Queensland, through eastern N.S.W. and Victoria to south-eastern South Australia, and in Tasmania. It is the most spectacular species of rice-flower at Kosciusko, being found just above the treeline in heaths. Reports of stockmen last century of a Kosciusko rose proved to be of this species (Maiden 1899). The large creamy flower heads have a superficial resemblance to a banksia rose.

Reference: Maiden (1899).

P. alpina F. Muell. ex Meissn. in DC., Prodr. 14: 511 (1857)
'Alpine rice-flower'

Low much-branched ± spreading prostrate to decumbent perennial usually less than 30 cm high, glabrous except the inflorescence; *stems* tough and wiry, often rooting towards the base, the branchlets ascending and rough with conspicuous leaf scars; *leaves* opposite, 5–10 × 2–2.5 mm, narrowly elliptic, subobtuse, concave above, decussate and usually crowded towards the ends of the branchlets; *petiole* ± 0.5 mm long; *involucral bracts* like the leaves but slightly wider, glabrous or with sparse appressed hairs on the midrib; *flowers* hermaphrodite or female, apparently functionally dioecious or almost so, pink to creamy white, antrorsely hairy, sessile on a densely hairy receptacle, in small terminal heads about 10–15 mm diam.; *hermaphrodite flowers: floral tube* 4.5–6 mm long; *calyx lobes* ± 2 mm long; *stamens* 2, shortly exserted; *anthers* ± 0.5–0.6 mm long, orange-yellow; *style* included or occasionally shortly exserted, the stigma small; *female flowers:* usually slightly smaller than the hermaphrodite flowers, the floral tube tardily circumciss above the ovary; *stamens* small and sterile; *style* exserted with a prominent stigma; *fruit* ovoid, 3–4 × ± 2 mm, enclosed in the persistent base of the floral tube.

[211–12]

Alpine and subalpine tracts of the Australian Alps.

Distribution *Notes and* *Habitat*

The breeding system in *Pimelea* is frequently complex and needs investigating in this as well as many other Australian species (see Burrows 1960). Alpine Rice-flower is frequently found sprawling over low rocks and boulders in heath, or growing nearby among grasses in tall alpine herbfield or sod tussock grassland. The flowers are sweetly perfumed.

Reference: Burrows (1960).

Myrtaceae

Kunzea Reichenb.

K. muelleri Benth., Flor. Aust. 3: 113 (1867)   'Yellow kunzea'

[213–14]   Erect or spreading shrub ± 0.5(–1) m high; *leaves* crowded, suberect 3–7 × ± 0.5 mm almost terete, the margins involute, with deep adaxial groove, minutely gland-dotted, villous becoming glabrescent with age, sometimes with a persistent line of hairs along the groove; *flowers* sessile, usually in small dense terminal or subterminal clusters subtended by persistent pale brown ovate villous bracts; *floral tube* villous, ± 3.5 mm long; *sepals* 5, triangular, ± 1.5 × 1.5 mm, villous, persistent, yellowish becoming brown and glabrescent with age; *petals* 5, yellow, orbicular-obovate, alternating with and slightly longer than the sepals; *stamens* numerous, conspicuous and much exceeding the petals; *style* glabrous, ± 5–7 mm long, often persistent; *capsule* 2–3-celled, deeply sunken in the persistent floral tube.

Distribution   Alpine, subalpine and montane tracts of Australian Alps and associated ranges, as far north as the A.C.T.

Notes and   This attractive yellow-flowered shrub forms extensive dwarf heaths in situations
Habitat   intermediate in exposure between the taller communities of *Oxylobium–Podocarpus* heath and the most wind-exposed *Epacris–Chionohebe* feldmark. It is also common in the wet *Epacris–Kunzea* heaths surrounding bogs.

Baeckea L.

B. gunniana Schau. in Walp., Repert. Bot. Syst. 2: 920 (1843)
'Alpine baeckea'

Glabrous densely branched aromatic shrub 0.5–1 m high (up to 2 m at lower altitudes), sometimes prostrate or spreading over rocks and boulders; *leaves* opposite, small and crowded, shortly petiolate, ± 2–4 × 0.6–0.8 mm, obovate-oblong, blunt, ± channelled above and convex but not keeled on the underside, covered with conspicuous oil-dots (visible under a 10× lens); *petioles* up to 0.5 mm long; *flowers* small and numerous, ± 4–5 mm diam., solitary in the upper axils on short pedicels ± 1 mm long; *bracteoles* linear, ± 1 mm long, soon caducous; *calyx lobes* 5, triangular, ± 0.5 mm long; *petals* 5, white, orbicular, 1–1.5 mm diam.; *stamens* usually 5 or sometimes up to 7, incurved, ± 0.5 mm long; *capsules* cup-shaped, ± 2 mm diam., the calyx lobes persistent.

[215]

Alpine, subalpine and montane tracts of south-eastern Australia, including Tasmania.

Distribution

Common in bogs and in wet rocky sites and along stream banks as a component of heaths.

Notes and Habitat

Reference: Willis (1967).

B. utilis F. Muell. ex Miq., Ned. Kruidk. Arch. 4: 150 (1856)
'Mountain baeckea'

Superficially similar to the preceding in general appearance but larger in all its parts, and differing mainly as follows: *leaves* ± flat above, keeled on the underside, 4–10 × 1–2.5 mm, elliptical to oblanceolate; *petioles* 1–1.5 mm long; *flowers* to 7 mm diam., on longer pedicels ± 2–3 mm long; *petals* ± 2 mm diam.; *stamens* usually 8(7–10).

[216–17]

Alpine, subalpine and montane tracts of south-eastern N.S.W. as far north as the Blue Mountains, and eastern Victoria where it descends to about 100 m above sea level in East Gippsland (Willis 1967).

Distribution

Common in wet rocky sites and along streams as a component of heaths, but less common in bogs (cf. *B. gunniana*). Plants near the summit of Mt Kosciusko are sometimes quite prostrate, but at lower altitudes along streams in the montane tract the species may grow to a height of 3 m.

Notes and Habitat

Reference: Willis (1967).

Onagraceae

Epilobium L.

E. tasmanicum Hausskn., Monogr. Epilob. 296, t. 20, f. 84 (1884)
'Snow willow-herb'

[218–19]

Prostrate perennial herb usually forming dense mats about 10–20(–60) cm diam.; *stems* creeping and rooting at the nodes, the tips ascending, glabrous or with two inconspicuous lines of pubescence decurrent from the margins of the petioles; *leaves* glabrous, shining, mostly opposite except alternate in the inflorescence, rather crowded, shortly petiolate or subsessile, ± 5–10(–15) × 2–4 mm, the blade narrowly elliptic or elliptic, obtuse or subobtuse, attenuate at the base, with 3–5 small distant glandular teeth on each side or subentire; *flowers* ± 8 mm diam., few in the upper axils; *sepals* 4, glabrous, narrowly triangular, 3–4 mm long, pale green usually tinged with pink on the margins; *petals* 4, white, shortly exceeding the sepals, deeply emarginate; *stamens* 8, alternately long and short; *anthers* yellow, 0.4–0.45 mm long; *stigma* clavate, subequal in length to the style; *capsule* glabrous, 0.8–2.5 cm long, green becoming reddish at maturity, on a pedicel 0.5–5 cm long; *seeds* pale yellowish brown, micropapillate, ± 1 × 0.4–0.5 mm, narrowly obovoid to obovoid, sharp-pointed; *coma* white, 3–5 mm long, caducous; $n = 18^\star$.

Distribution

Alpine tract of the Kosciusko area, N.S.W., and high mountains of Tasmania and New Zealand (South Island).

Notes and Habitat

Locally common below snow patches as a characteristic component of *Coprosma–Colobanthus* feldmark, short alpine herbfield and colonizing disturbed areas and erosion pavements. This species is superficially similar to *E. curtisiae* Raven which is found in the subalpine tract, but the latter is readily distinguished by the seeds which lack a coma of hairs and have a pale inflated cellular rim around the lower margin, pale rose-purple flowers, slightly thicker capsules which are subsessile or very shortly stalked, and the usually undulate margins of the leaves.

References: Raven (1963); Raven and Raven (1976)*.

E. gunnianum Hausskn., Oest. Bot. Z. 29: 149 (1879)
'Gunn's willow-herb'

Erect perennial herb about 10–40(–70) cm high with numerous leafy stolons; [220–1]
stems branched and creeping at the base, often reddish, puberulent with short
curved hairs, with raised lines decurrent from the margins of the petioles; *leaves*
mostly glabrous but with short curved hairs on the margins and the midvein
below, subsessile or shortly petiolate, the stem leaves mostly opposite, becoming
alternate above and in the inflorescence, 1.5–4(–6) × 0.5–1.5(–2) cm, narrowly
elliptic, elliptic or ovate, subcordate or narrowed at the base, obtuse to subacute,
the margins closely serrulate or erose-denticulate and usually undulate; *flowers*
solitary in the upper axils ± 1–1.8 cm diam., the short floral tube with a conspicuous fringe of white hairs inside near the base; *sepals* 4, narrowly triangular,
3.5–5.5(–7) mm long, often reddish-tinged and conspicuously rugose distally,
puberulent with a mixture of curved and glandular hairs; *petals* 4, broadly obovate, deeply emarginate, purplish red to mauve or almost white; *stamens* 8,
alternately long and short; *stigma* broadly clavate-capitate, exceeding the anthers
at maturity; *capsule* puberulent, (3.5–)5–6(–7.5) cm long, on a pedicel 0.3–1.5
cm long; *seeds* narrowly obovoid, pale brown, ± 1–1.5 × 0.4–0.5 mm,
micropapillate to smooth, with a pale cellular rim around the margin on the
adaxial side which projects distally as a short obtuse beak to which the coma
of hairs is attached; *coma* white, (4.5–)5–7 mm long; $n = 18^*$.

Occasional on the Northern and Central Tablelands and South Coast of New *Distribution*
South Wales and the extreme north-eastern coastal area of Victoria, more common on the Southern Tablelands of New South Wales, the Australian Alps and
associated ranges and the Eastern Highlands of Victoria, Tasmania, and from
two localities on the west coast of Nelson, New Zealand.

This is the most common *Epilobium* above the treeline, found in moist sites *Notes and*
in several communities including tall alpine herbfields, sod tussock grassland, *Habitat*
bogs and heath. It forms extensive colonies in wet seepage areas around the
bases of large boulders near bogs or along stream banks. Raven (1973) suggests
that the New Zealand localities of this species might possibly be due to long-
distance dispersal from Australia.

References: Raven (1973); Raven and Raven (1976)*.

E. sarmentaceum Hausskn., Oest. Bot. Z. 29: 149 (1879)
'Mountain willow-herb'

Perennial herb 10–20(–40) cm high from loosely rhizomatous base, often pro- [222–3]
ducing short leafy stolons; *stems* erect or ascending, pubescent with short curved
hairs; *leaves* glabrous except for short curved hairs on the margins and on the
mid-rib proximally, shortly petiolate, mostly opposite, alternate in the inflorescence, 1–4(–5) × 0.5–1.5 cm, lanceolate to elliptic, distantly toothed with
3–8 teeth on each side, acute or subobtuse; *flowers* solitary in the upper axils,
1–1.3 cm diam., the short floral tube glabrous within; *sepals* 4, narrowly triangular,
puberulent, ± 3–4(–4.5) mm long; *petals* 4, broadly obovate, deeply emarginate,
mauve, purplish pink or almost white; *stamens* 8, alternately long and short;
stigma clavate, subequal in length to the style; *capsule* puberulent, (2–)3–5 cm

long, on a pedicel (1–)1.5–5 cm long; *seeds* narrowly obovoid, pale brown, ±
1.2–1.4 × 0.5–0.6 mm , micropapillate; *coma* white, 5–8 mm long; $n = 18^*$.

Distribution   Alpine and subalpine tracts of the Australian Alps and mountains of Tasmania.

Notes and   This species is much less common than *E. gunnianum* and is found in drier sites,
Habitat   usually in crevices in large rock outcrops in tall alpine herbfields and sod tussock
grassland.

Reference: Raven and Raven (1976)*.

Haloragaceae

KEY   1   Leaves linear-terete or subulate (rarely the lowermost with a few filiform
segments); fruit splitting into 4 mericarps at maturity
Myriophyllum pedunculatum (p. 198)

1   Leaves not as above, the margins crenate-serrate; fruit indehiscent
Gonocarpus (p.199)

Myriophyllum L.

M. pedunculatum Hook. f., Hook. Lond. J. Bot. 6: 474 (1847)
'Mat water-milfoil'

[224–5]   Glabrous mat-forming perennial herb, dwarf and only a few cm at high altitudes
(to 10 cm or more high at lower elevations); *stems* weak and slender, slightly
fleshy, decumbent or ascending, creeping and rooting at the lower nodes; *leaves*
sessile, opposite, mostly entire, linear-terete to subulate, 2–6(–10) × 0.5–1 mm,
with a small dark thickening at the apex, sometimes the lower submerged leaves
divided into a few filiform segments; *flowers* small, unisexual, solitary in the
upper axils with the male above and the female below, or the plants dioecious;
male flowers: pedicels 2–4(–10) mm long, with a pair of small linear bracteoles
near the base; *sepals* 4, erect, triangular, 0.7–1 mm long; *petals* 4, reddish, hooded,
oblong or obovate in outline, ± 2–2.5 mm long, recurved or reflexed at anthesis;
stamens 8; *anthers* 1–1.8 mm long, the filaments sometimes elongating to ± 1
cm at maturity; *female flowers:* subsessile, subtended by 2 small bracteoles; *sepals*
minute; *petals* absent; *stamens* absent; *stigmas* feathery, recurved; *mericarps* 4, connate, up to 1 mm long and minutely tuberculate in fruit.

Distribution   Tablelands and ranges of eastern N.S.W., Victoria and Tasmania; also New
Zealand, Stewart and Chatham Islands and New Guinea.

Notes and   Forms extensive turf-like areas on the margins of lakes, pools and bogs, often
Habitat   submerged in several inches of water, and in shallow depressions in fen subject
to periodic inundation, frequently assuming a reddish hue as the pools dry out.

Gonocarpus Thunb.

G. micranthus Thunb., Nov. Gen. Plant. 55 (1783)

subsp. *micranthus*   'Creeping raspwort'

Prostrate or procumbent creeping perennial herb; *stems* up to 30 cm long, glabrous   [226]
or almost so, intricately branched, rooting at the nodes, ± ascending at the
tips; *leaves* opposite, decussate, shortly petiolate, ± 3–5(–8) × 3–4(–6) mm,
broadly ovate, ovate-cordate or suborbicular, obtuse or rounded at the apex,
the margins slightly thickened, crenate-serrate; *petioles* ± 0.5–1 mm long; *inflorescence* slender, not or scarcely branched, often only a few cm long; *bracts* deciduous, 0.5–0.8 mm long; *flowers* small, 0.8–1 mm long, reddish, on very short
pedicels; *sepals* 4, persistent, ovate-triangular, ± saccate at the base, ± 0.5 mm
long; *petals* 4, caducous, boat-shaped, 0.5–1 mm long; *stamens* 8, the filaments
up to 0.7 mm long; *anthers* ± 0.5–0.7(–0.9) mm long; *stigmas* feathery; *fruit*
grey or reddish, glabrous, smooth and shining, obovoid to subglobose, longitudinally 8-ribbed, ± 0.5–1 × 0.5–0.7 mm (excl. the persistent sepals), nodding
at maturity.

N. Queensland (Bathurst Harbour), ranges of SE. Queensland, eastern New   *Distribution*
South Wales, Victoria, south-eastern South Australia and Tasmania; also moun-
tainous areas of India, south-east Asia, Japan, the Philippines, Malesia, New
Guinea and New Zealand.

Common in wet sites, as in fen and bogs, pools in sod tussock grassland and   *Notes and*
on the margins of creeks, usually growing in moss or forming mats on wet mud.   *Habitat*
The subspecies *ramosissimus* Orchard is a larger, ± erect form with a much-
branched, open inflorescence and is found on the Great Dividing Range from
central Queensland to central N.S.W. and adjacent coastal areas.
Reference: Orchard (1975).

G. montanus (Hook. f.) Orchard, Bull. Auckland Inst. Mus. No. 10: 172 (1975)

Prostrate to ascending perennial herb to about 40 cm diam. from erect woody   [227]
rootstock; *stems* thin and wiry, 4-angled, rooting towards the base, antrorsely
appressed-hairy on opposite sides (above the leaves); *leaves* opposite, decussate,

shortly petiolate, coriaceous, ± 5–10 × 2–6 mm, ovate to broadly lanceolate, acute to subacute, the margin thickened, distantly crenate-serrate, glabrous or with sparse hairs proximally on the margins and the midrib below; *petioles* 0.5–1 mm long; *flowers* small, 4-merous, solitary in the axils of small leaf-like bracts, in terminal or subterminal racemes, subsessile or on short pedicels ± 0.5–1.5 mm long, subtended by 2 small ciliolate bracteoles 0.5–1 mm long; *sepals* broadly triangular, 0.5–0.7 mm long; *petals* herbaceous, caducous, ± 1–1.5 mm long, boat-shaped, ciliolate on the keel; *stamens* 4; *anthers* 0.5–0.8 mm long, shortly exserted; *stigmas* sessile, feathery; *fruit* greyish, broadly ellipsoid to subglobular, with eight longitudinal ribs and a few ± obscure oblique oblong tubercles between the ribs, 1–1.5 × 1–1.2 mm (excl. the persistent sepals), micropapillate, glabrous or with sparse minute hairs on the ribs and tubercles.

Distribution   Alpine and subalpine tracts of the Australian Alps and associated ranges as far north as the A.C.T., and mountains of Tasmania and New Zealand.

Notes etc.   Rather sparsely distributed in tall alpine herbfields, usually in stonier sites.

Umbelliferae

KEY
1	Leaves compound	2
1	Leaves simple	4

2	Umbels simple	*Oreomyrrhis* (p. 201)
2	Umbels compound	3

3   Leaves stiff, conspicuously jointed, the ultimate segments linear, pungent-pointed   *Aciphylla glacialis* (p. 204)
3   Leaves not as above   *Gingidia algens* (p. 205)

4	Leaves long and narrow	5
4	Leaves with laminae ± suborbicular in outline	6

5   Leaves entire, transversely septate   *Aciphylla simplicifolia* (p. 203)
5   Leaves toothed or lobed at the apex, not septate

                                                                Oschatzia cuneifolia (p. 205)

6   Leaves not cordate at the base   *Dichosciadium ranunculaceum* (p. 207)
6   Leaves cordate at the base   7

7   Leaves thick and slightly fleshy, entire or obscurely crenulate; scapigerous herb   *Diplaspis hydrocotyle* (p. 206)
7   Leaves thin, distinctly lobed and crenate; creeping herb forming dense patches   *Schizeilema fragoseum* (p. 207)

Oreomyrrhis Endl.

1 Involucral bracts and leaves glabrous except ciliate and/or ciliolate on the margins 2
1 Involucral bracts and leaves hairy 3

2 Leaf sheaths conspicuously ciliate; stylopodium and style together ± 1 mm long; small mat-forming herb *O. pulvinifica* (p. 201)
2 Leaf sheaths glabrous; stylopodium and style together up to 0.5 mm long; tufted rosette herb *O. ciliata* (p. 202)

3 Strongly caulescent and umbellately branched above; peduncles slender, 1–5 cm long, ± hidden among the upper leaves; fruiting pedicels shorter than the involucre; involucral bracts ± velutinous with short spreading hairs *O. brevipes* (p. 202)
3 Acaulescent, or caulescent towards the base; peduncles stout and conspicuous, usually much overtopping the leaves in fruit; fruiting pedicels mostly longer than the involucre; involucral bracts hirsute with antrorsely appressed or ascending hairs *O. eriopoda* (p. 203)

Mathias and Constance (1955) suggest that introgression occurs between *O. brevipes* and *O. eriopoda* in the Kosciusko area.

O. pulvinifica F. Muell., Fragm. Phyt. Aust. 8: 185 (1874)
'Cushion carraway'

Small mat-forming perennial herb 2–10 cm high from a much-branched shortly creeping rootstock; *leaves* pinnate, the blades oblong-oval in outline, 0.5–2 × 0.4–1.5 cm, the pinnae sessile or shortly stalked, the lower pairs ternate or pinnatisect and the upper entire, or all pinnatisect, the ultimate segments linear, subacute, glabrous or the margins ciliolate; *petioles* subequal to or longer than the blades, with broad, scarious, conspicuously ciliate sheathing bases; *peduncles* several, hirsute with ± spreading or loosely retrorse hairs, subequal to or shortly overtopping the leaves in fruit; *involucre* of 5–8 oval or oblong, usually entire obtuse bracts, united at the base, glabrous except the margins ciliolate; *flowers* small, the petals white (usually drying yellow), glabrous; *pedicels* up to 7 mm long and usually shortly exceeding the involucre in fruit, hirsute with spreading or loosely antrorse hairs; *fruit* narrowly ovate-oblong, 2.5–3(–4) × 1–1.5 mm, the mericarps 5-ribbed, the stylopodium and style together up to 1 mm long.

[228–9]

Australian Alps, mostly above the treeline.

Distribution

A characteristic component of short alpine herbfield, also in damp sites in tall alpine herbfield. Well-grown plants of this species sometimes superficially resemble small specimens of *O. ciliata*, but can be readily distinguished from the latter by the ciliate leaf sheaths.

Notes and Habitat

Reference: Mathias and Constance (1955).

O. ciliata Hook. f., Hook. Lond. J. Bot. 6: 471 (1847)

[230] Tufted perennial herb (2–)15–30(–45) cm high, acaulescent or almost so, from a slender tap-root; *leaves* in a basal rosette, glabrous except the ultimate segments ciliolate, the blades pinnate, oblong in outline, 1–9 × 0.5–3 cm, the pinnae sessile, deeply pinnately lobed or pinnatisect with linear-lanceolate, acuminate, mucronate segments; *petioles* shorter or longer than the blades, with broad scarious, usually reddish purple sheathing bases; *peduncles* 2–6, densely hirsute above with retrorse or spreading hairs, elongating and usually overtopping the leaves in fruit; *involucre* of 6–8 ovate or ovate-lanceolate, acute, usually entire bracts, united at the base, glabrous except for the ciliolate margins; *flowers* small, the petals white, glabrous or sparsely pubescent on the back; *pedicels* very unequal, to 15 mm long, exceeding the involucre in fruit, hirsute with antrorse hairs; *fruit* ovoid to oblong-ovoid, 2–5 × 1–2 mm, the mericarps 5-ribbed, the stylopodium and style together up to 0.5 mm long.

Distribution   Barrington Tops, N.S.W., alpine and subalpine tracts of the Australian Alps and associated ranges as far north as the A.C.T., and mountains of Tasmania.

Notes and   This species is very common around subalpine bogs and is locally common in
Habitat   bogs above the treeline; it has a very penetrating sickly-sweet odour which often persists long after specimens have been pressed and dried. We suggest that an appropriate common name for this species might be Bog Carraway.

Reference: Mathias and Constance (1955).

O. brevipes Mathias & Constance, Univ. Calif. Publ. Bot. 27(6): 390 (1955)

[231–2] Rounded caulescent perennial herb 10–25(–40) cm high from a stout tap-root, the base of the plant covered with the remains of old sheaths; *stems* numerous, softly hirsute, leafy, umbellately branched above; *leaves* ± velutinous, the blades pinnate, oval-oblong in outline, 2–7.5 × 1.5–4 cm, the pinnae once or twice pinnatisect with linear acute segments; *petioles* softly hirsute, longer than the blade, sheathing at the base; cauline leaves similar to the basal leaves but smaller; *peduncles* short, slender, 1–5 cm long, softly hirsute with spreading hairs, numerous in the upper axils and not or scarcely overtopping the leaves; *involucre* of ± 8 narrowly oblong to oval, entire or lobed bracts, united at the base, ± velutinous on both sides; *flowers* small, the petals white to deep magenta, shortly hirsute on the back; *ovary* glabrous or with sparse hairs; *pedicels* 1–5 mm long, shorter than the involucre, softly hirsute with short spreading hairs; *fruit* glabrous, oblong-ovoid, 5–6 × 2–2.5 mm, the mericarps 5-ribbed, dark brown or almost black at maturity.

Distribution   Alpine tract of the Australian Alps from the Kosciusko area in N.S.W. to the Cobberas mountains, Mt Nelse and Mt Skene in Victoria.

Notes and   See Notes under *O. eriopoda*. *O. brevipes* is found mainly in rocky areas, often
Habitat   on and around the bases of large granitic tors or boulders in tall alpine herbfield, and also in *Epacris–Chionohebe* feldmark; Mathias and Constance (1955) indicate its resemblance to the Tasmanian endemic, *O. sessiliflora* Hook.f. We suggest that an appropriate common name for this species might be Rock Carraway.

Reference: Cerceau-Larrival (1974).

O. eriopoda (DC.) Hook. f., Flor. Tas. 1: 162 (1856)
'Australian carraway'

Perennial herb 5–30(–50) cm high, acaulescent or branched towards the base [233]
(see Notes below), from a stout or slender tap-root; *leaves* pinnate, shortly hirsute
with appressed or ascending hairs, the blades oblong-ovate in outline, 2.5–15
× 1–8 cm, the pinnae once or twice pinnatisect with oblong subacute ultimate
segments; *petioles* hairy, usually longer than the blade, sheathing at the base;
peduncles stout, long and conspicuous and usually much overtopping the leaves
in fruit, hirsute with retrorse or spreading hairs; *involucre* of 8–12 narrowly oblong,
oblanceolate or ovate bracts, entire or toothed or lobed distally, united at the
base, with short appressed or ascending hairs on both sides; *flowers* small, the
petals white or tinged with pink, hirsute on the back; *pedicels* very unequal,
5–15 mm long, usually longer than the involucre, hirsute with antrorse or spreading hairs (see Notes below); *fruit* 3.5–6.5 × 1.5–2 mm, narrowly ovate-oblong,
the mericarps 5-ribbed.

Mountains and tablelands of eastern N.S.W., the A.C.T., Victoria, the Mt Lofty *Distribution*
Range in South Australia, and Tasmania.

Common above treeline, in tall alpine herbfields and heaths. Plants from lower *Notes and*
altitude tend to be slender and more strictly acaulescent, with appressed or *Habitat*
subappressed pubescence, retrorse on the long peduncles and antrorse on the
pedicels which are much longer than the involucral bracts. At higher altitudes,
however, and especially above the treeline, the plants tend to become more
robust, to show a greater degree of branching, and the peduncles and pedicels
to be shorter with more copious and spreading pubescence. Mathias and Constance (1955) have pointed out that this suggests introgressive hybridization
with *O. brevipes*.

Reference: Mathias and Constance (1955).

Aciphylla J. R. & G. Forst.

---

1. Leaves simple, grass-like, transversely septate, not pungent KEY TO THE
        A. simplicifolia (p. 203) SPECIES
1 Leaves compound, the segments conspicuously jointed, the ultimate segments
   stiff and pungent-pointed; more robust than the preceding
        A. glacialis (p. 204)

A. simplicifolia (F. Muell.) Benth., Flor. Aust. 3: 375 (1867)
'Mountain aciphyll'

Glabrous tufted dioecious perennial herb 15–40(–50) cm high, the base of the [234]
plant surrounded by the remains of old leaves; *stems* erect or ascending, often

purple-tinged, the inflorescence overtopping the leaves; *leaves* radical, simple, subcylindrical, linear to very narrowly oblanceolate, obtuse, conspicuously transversely septate, deeply grooved on the upper surface, (10–) 15–25 (–30) cm long excluding the broad sheathing bases; *inflorescence* with a terminal compound umbel and alternate or subopposite lateral compound umbels, the male inflorescence open and delicate, the female narrower with stouter peduncles; *bracts* mostly shorter than the compound umbels, the lowermost usually with a short rudimentary lamina, the upper reduced to the sheaths; *male flowers:* ± 1.5–2 mm diam.; *calyx teeth* 5, minute; *petals* creamy white or pink-tinged, ± 0.7–1 × 0.5 mm, obovate, slightly incurved at the apex; *stamens* 5, subequal to or shorter than the petals; *ovary* and styles rudimentary; *female flowers: stamens* reduced and sterile; *fruit* dorsally compressed, narrowly oblong-elliptic in outline, ± 8–10 × 3 mm, the mericarps with 2–3 very narrowly winged dorsal ribs and two slightly more broadly winged lateral ribs.

Distribution   Alpine and subalpine tracts of the Australian Alps and associated ranges as far north as the A.C.T.

Notes and Habitat   The genus is best developed in New Zealand where there are about 37 species. This species is readily distinguished from the more robust *A. glacialis* because of its simple leaves and less conspicuous inflorescence, and although now fairly common in sod tussock grassland and tall alpine herbfields, like the latter species it was greatly reduced in abundance by earlier livestock grazing.
References: Dawson (1971); Oliver (1956).

A. glacialis (F. Muell.) Benth., Flor. Aust. 3: 375 (1867)
'Mountain celery; snow aciphyll'

[235–8]   Robust rather sprawling dioecious glabrous perennial ± 30–70 cm high from a tough erect or oblique branched rootstock, the base of the plants thickly surrounded by the remains of old leaves; *stems* erect or ascending, sometimes shortly decumbent at the base, the inflorescence overtopping the leaves; *leaves* dark green, leathery, crowded at the base of the plant in dense fan-shaped clusters, 10–25 cm long including the very broad fleshy imbricate sheathing bases, the blades ovate or elliptic in outline, 2(–3)-pinnate, conspicuously jointed, the ultimate segments linear, suberect, stiff, channelled, ± 2–4 cm long, pungent-pointed; *inflorescence* with a terminal compound umbel and alternate, subopposite or ± whorled lateral compound umbels, the male inflorescence spreading, the female more robust and contracted with thicker peduncles; *bracts* shorter than the compound umbels, the lower bracts leaf-like, the upper reduced to the sheaths; *male flowers:* ± 2.5–3 mm diam., the ovary and styles rudimentary; *calyx teeth* 5, minute; *petals* creamy white, obovate, 1–1.5 mm long, slightly incurved at the apex; *stamens* 5, subequal to or slightly exceeding the petals; *female flowers: stamens* reduced and sterile; *fruit* dorsally compressed, oblong-elliptic in outline, 5–6 × 2.5–3.5 mm, the mericarps with 2–3 very narrowly winged dorsal ribs and 2 slightly more broadly winged lateral ribs.

Distribution   Alpine and upper subalpine tracts of the Australian Alps.

Notes and Habitat   This showy species, with its large white male inflorescences and prickly dissected leaves forms extensive communities in tall alpine herbfields. As already noted

(Costin 1958), it is palatable to livestock and has become much more common since grazing ceased; it is still increasing in abundance at Kosciusko.
Reference: Costin (1958).

Gingidia Dawson

G. algens (F. Muell.) Dawson, Contrib. Herb. Aust. No. 23: 1 (1976)

Glabrous erect or ascending perennial herb ± 15–30(–45) cm high from a long ± branched erect or ascending rootstock; *leaves* mostly radical, petiolate, simply pinnate, ± 10–25 cm long (incl. the petiole); *segments* sessile or the lowermost stalked, rhomboid or obovate-cuneate in outline, ± 1–4 cm long, variably lobed or lacerate or the lower deeply pinnatifid, the lobes incised or acutely toothed distally; *petioles* expanded and broadly sheathing at the base; *stipules* absent; cauline leaves few, reduced, with narrower segments; *umbels* compound, the primary rays 4–7, unequal; *involucral bracts* 1–3, free, filiform; *flowers* small, white, ± 2–2.5 mm diam.; *calyx teeth* small and inconspicuous; *petals* 5, white, ± 1 mm diam., suborbicular, abruptly constricted at the apex to a linear-acuminate inflexed tip; *stamens* 5, subequal in length to the petals; *anthers* pink, 0.5 mm long; *fruit* glabrous, ovoid-oblong ± 5 × 2.5 mm, the mericarps subequally 5-ribbed. [239–40]

Endemic to the alpine and subalpine tracts of the Kosciusko area, N.S.W., as far north as Mt Jagungal. *Distribution*

Although fairly common in the subalpine tract, this species is rather uncommon above treeline in moist tall alpine herbfield sites marginal to bogs. It has a strong aniseed odour when crushed. The closely related *G. harveyana* (F. Muell.) Dawson which is more common and widespread in the subalpine tract, can be readily distinguished by the leaves, the segments of which are linear or linear-lanceolate and entire, or the lower segments may be ± divided into linear or linear-lanceolate lobes. *Notes and Habitat*

References: Cerceau-Larrival (1974); Dawson (1961, 1967, 1974, 1976); Moar (1966).

Oschatzia Walp.

O. cuneifolia (F. Muell.) Drude, Natürl. Pflanzenfam. 111 (8): 128 (1897)   'Wedge oschatzia'

Glabrous perennial herb to 20 cm high from a loosely rhizomatous base; *flowering stems* slender, leafless, erect or ascending; *leaves* radical, 3–12 × 0.4–1 cm, narrowly cuneate or subspathulate, attenuate to a long petiole, with 3–9 irregular acute teeth or lobes at the apex; *petiole* shortly sheathing and often purple-tinted [241] at the base; *flowers* ± 4–5 mm diam., 2–9 in a loose open irregularly compound umbel; *calyx teeth* inconspicuous; *petals* 5, white, ± 2 × 1 mm, oval or ovate, slightly imbricate, deciduous; *stamens* 5, alternate with and shorter than the petals; *fruit* ± 3.5 × 3 mm (excl. stylopodium), ovoid, scarcely compressed; *mericarps* equally 5-ribbed.

Distribution   Alpine and subalpine tracts of the Australian Alps.

Notes and Habitat   Sparsely distributed in damp sites in tall alpine herbfields and in bogs. A small genus of two species, one in the Australian Alps and one in Tasmania.

Reference: Cerceau-Larrival (1974).

Diplaspis Hook. f.

D. hydrocotyle Hook. f., Hook. Lond. J. Bot. 6: 469 (1847)
'Stiff diplaspis'

[242]   Small perennial herb 2–12 cm high from horizontal or ascending fleshy rhizomes 3–4 mm diam.; *leaves* rosulate, long-petiolate, thick and slightly fleshy, ± 0.5–1.5 × 0.5–1.5 cm, ovate to suborbicular in outline, cordate at the base, glabrous or with very sparse coarse erect hairs on the upper surface, paler on the underside, the margins slightly revolute, entire or obscurely crenate; *petioles* 1–7 cm long, glabrous or with sparse stiff spreading hairs, with broad membranous sheathing bases, the sheaths with small entire, dissected or fimbriate filiform points at the summit; *scape* simple, erect, glabrous or with sparse stiff spreading hairs, becoming elongated, hollow and up to 5 mm diam. in fruit; *bracts* narrow-oblong, glabrous or sparsely fimbriate ± 4–5 mm long; *flowers* small and inconspicuous, ± 2.5 mm diam., about 10–20 in a simple umbel; *calyx* rudimentary; *petals* 5, imbricate at the base, yellowish green or tinged with red, broadly ovate, obtuse, 1–1.3 mm long; *stamens* 5, alternate with and shorter than the petals; *fruit* dorsally compressed, the flat outer faces of the mericarps broadly ovate or oval in outline, ± 3 × 2–2.4 mm (excl. stylopodium); *carpophore* persistent, shortly bifid at the apex.

Distribution   Alpine and subalpine tracts of the Australian Alps and high mountains of Tasmania.

Notes and Habitat   Locally common in short alpine herbfield and bogs; also damp sites in tall alpine herbfields. A small genus of two species, closely related to the genus *Huanaca* of South America (Curtis 1963 and Mathias and Constance 1971).

References: Cerceau-Larrival (1974); Curtis (1963); Mathias and Constance (1971).

Schizeilema Domin

S. fragoseum (F. Muell.) Domin, Bot. Jb. 40: 584 (1908)

'Alpine pennywort'

Glabrous perennial herb from slender, branched, ± matted horizontal or ascending rhizomes, forming dense dark green patches up to about 12 cm high and 0.5 m diam.; *flowering stems* slender, ascending, hidden among the leaves; *radical leaves* on slender petioles up to 10 cm long, the blade ± 1–2 cm diam., orbicular-cordate to reniform, with 5–9 irregular ± crenate lobes, paler on the underside; *stipules* membranous, partially adnate to the petiole, acuminate or laciniate distally; *umbels* simple, 4–8-flowered, sessile or apparently pedunculate; *involucral bracts* 5–8, ovate-lanceolate, shortly connate at the base, lobed and/or laciniate distally; *flowers* small, 5-merous; *calyx teeth* persistent, ovate-triangular, up to half as long as the petals; *petals* oval, ± 1 mm long, pale greenish yellow, caducous; *fruit* smooth, slightly dorsally compressed, ± 2–2.5 × 1.3–1.5 mm, ovate-elliptic in outline, deeply constricted between the mericarps which are oblong-elliptic in cross-section and attached by their broad faces; *carpophore* absent. [243]

Endemic to the alpine and subalpine tracts of the Australian Alps. *Distribution*

Fairly common in tall alpine herbfields, especially in moist shady sites often around the bases of rocks. This species looks very like a *Hydrocotyle* but differs in the dorsally compressed fruit, etc. *Notes and Habitat*

Reference: Cerceau-Larrival (1974).

Dichosciadium Domin

D. ranunculaceum (F. Muell.) Domin, Feddes Repert. 5: 104 (1908) var. *ranunculaceum*

Acaulescent perennial herb with a thick simple or branched fleshy tap-root; *leaves* long-petiolate, numerous in a dense radical rosette about 10–20 cm diam., ± 1.5–3 × 1.5–3.5 cm, in outline broadly obovate-cuneate to suborbicular with broadly cuneate, subtruncate or rounded base, palmately 3–5 lobed, the lobes ± deeply and irregularly incised-crenate, sparsely hispid along the veins and margins or almost glabrous, sometimes with more conspicuous coarse appressed hairs on the underside of the lobes; *petioles* up to 10 cm long, expanded at the base, coarsely hispid with stiff spreading hairs and small stellate and ± dendritic hairs; *peduncles* decumbent, usually shorter than the leaves, hispid like the petioles, each with an irregular usually simple umbel of 3–6 small flowers on uneven ± hispid pedicels; *bracts* unequal, narrowly oblong, blunt, ± united at the base, usually shorter than the longest pedicels; *flowers* white, 5–10 mm diam.; *sepals* 5, indistinguishable from the petals and deciduous with them; *petals* 5, ovate or elliptic, 2.5–3 × 1.5–2 mm; *stamens* 5, shorter than the petals and alternating with them; *anthers* purplish ± 0.8 mm long; *fruit* dorsally compressed, the meri- [244]

carps with the outer faces flattened and elliptic in outline, ± 5 × 3 mm (excl. stylopodium).

Distribution   Alpine tract of the Kosciusko area, N.S.W., with a variety in Tasmania.

Notes and Habitat   Often forming extensive patches in short alpine herbfield, and in bogs and damp sites in tall alpine herbfields. When fresh, the tap-root has a distinct parsnip-like odour.

Epacridaceae

KEY

1   Leaves broadly sheathing at the base; corolla circumciss near the base, the lobes not opening, the upper part operculate and falling entire
Richea continentis (p. 208)
1   Leaves not as above; corolla opening normally   2

2   Flowers in very small terminal or subterminal spikes, each flower subtended by a bract and two bracteoles; fruit drupaceous
Leucopogon montanus (p. 212)
2   Flowers solitary in the upper axils, each flower subtended by 4 or more small imbricate bracts   3

3   Corolla lobes glabrous; style inserted in a depression between the carpels; fruit a capsule opening by 5 valves   *Epacris* (p. 209)
3   Corolla lobes densely bearded; style terminal; fruit drupaceous
Pentachondra pumila (p. 212)

Richea R. Br.

R. continentis B. L. Burtt, Curtis Bot. Mag. 163, sub. t. 9632 (1941)   'Candle heath'

[245–6]   Much-branched subshrub to about 1 m high with decumbent woody stems, often freely rooting and creeping extensively through *Sphagnum* bogs; *branchlets* ascending with persistent old leaves at the base; *leaves* sessile, crowded, with broad chartaceous imbricate sheathing bases, the blades erect or spreading, narrowly triangular, stiff and pungent, 2–3.5 cm long and 0.4–0.7 cm wide at the base, the margins scaberulous; *inflorescence* terminal, spike-like, at first short and enclosed in caducous brown bracts, becoming elongated and interrupted at maturity, ±5–15 cm long on a reddish stalk about as long; *rachis* slightly flexuose, puberulent; *bracts* leaf-like on the stalk, becoming very short and broad upwards, those towards the apex with incurved or uncinate tips; *flowers* 5-merous, creamy white, subsessile, clustered on short puberulent lateral branches which are up to 5 mm long; *sepals* imbricate, persistent, broadly ovate, 2.5–3 mm long; *corolla* not opening, swollen and fleshy, ±5–8 × 4–5 mm, ovoid or obovoid, umbonate at the apex, circumciss near the base, the upper part operculate and falling entire; *stamens* with filaments 2–3 mm long, the anthers ±1 mm long; *ovary* 5-lobed, subtended by 5 ovate-triangular scales; *style* persistent, sunk in a depression in the ovary, ±1 mm long; *stigma* capitate; *capsule* loculicidal, ± 5 mm diam. when open, the persistent placentas raised on short recurved stalks around the base of the style; $2n = 26$ (as *R. gunnii*)★.

Alpine and subalpine tracts of the Australian Alps and associated ranges as far *Distribution* north as the A.C.T.

This species is a major component of bogs, where it is easily recognizable by *Notes and* its sharp-pointed sheathing leaves and showy spikes of creamy white flowers. *Habitat* It is closely related to *R. scoparia* Hook. f. of Tasmania.

Reference: Smith-White (1955)★.

Epacris Cav.

1	Leaves lanceolate to elliptic-lanceolate, ± 5–10 mm long; style shortly exserted, 4–5 mm long *E. paludosa* (p. 209)	KEY TO THE SPECIES
1	Leaves not as above, less than 5 mm long; style included, less than 2 mm long 2	

2	Leaves rhomboid to obovate-rhomboid; style longer than the ovary, extending about half way up the corolla tube; bracts and sepals strongly tinged with reddish brown; pollen grains papillate *E. glacialis* (p. 210)
2	Leaves not as above; style subequal to or shorter than the ovary; bracts and sepals pale stramineous or faintly tinged with pale brown; pollen grains smooth 3

3	Leaves petiolate, broadly ovate to broadly ovate-cordate *E. microphylla* (p. 211)
3	Leaves sessile, elliptic or ± oblong *E. petrophila* (p. 211)

E. paludosa R. Br., Prodr. Flor. Nov. Holl. 551 (1810)
'Swamp heath'

Shrub to 1 m or more high, erect or sometimes spreading over rocks in exposed [247–8] situations; *branchlets* puberulent, soon glabrescent; *leaves* thick, rigid, shortly petiolate ± 5–10 × 1.5–3 mm, lanceolate or elliptic-lanceolate, pungent-pointed, shining above, paler with the midrib prominent on the underside, the margins obscurely serrulate; *petiole* ± 0.5 mm long, articulate at the base; *flowers* 6–8 mm diam., shortly pedicellate, solitary in the axils, in short leafy clusters at the ends of the branchlets; *bracts* and *sepals* acute, usually ciliolate and tinged with reddish brown; *sepals* 4–5 mm long, ovate-or oblong-lanceolate; *corolla* white, the tube cylindrical, 4–5 mm long, subequal to or shortly exceeding the calyx, the lobes broadly ovate, obtuse, about 3 mm long, overlapping at the base; *stamens* attached about two-thirds up the tube; *anthers* 1.2–2 mm long, not exserted; *ovary* ± 1 mm long, subtended by 5 small scales; *style* 4.5 mm long, swollen about the middle, the large stigma ± flush with the top of the corolla tube.

Distribution   Central and southern tablelands and ranges of N.S.W. as far north as the Blue Mountains, the Victorian Alps and associated ranges, and Flinders Island.

Notes and   A characteristic shrub of bogs; also along rocky watercourses as a component
Habitat   of heaths.

E. glacialis (F. Muell.) M. Gray, Contrib. Herb. Aust. No. 26: 5 (1976)

[249–51]   Prostrate to decumbent shrub ± 5–30 cm high, the wiry stems freely rooting towards the base; *branchlets* obscurely puberulent, soon glabrescent; *leaves* coriaceous, subsessile or very shortly petiolate, ± 2–3.5 × 1.5–2.5 mm, rhomboid to obovate-rhomboid or broadly so, obtuse, flat or slightly concave above, keeled distally on the underside, the keel ± thickened at the slightly upturned apex, the margins scaberulous in the upper half and sometimes also proximally; *flowers* solitary and sessile in the upper axils, forming small clusters at the ends of the branchlets; *bracts* reddish brown, fimbriolate, the lower broadly ovate, the upper ovate-elliptic, obtuse; *sepals* ± 3.5–4.5 × 1.5–2 mm, ovate-elliptic to elliptic, obtuse, fimbriolate, tinged with reddish brown distally; *corolla* white, ± 1 cm diam., thickened in the throat, the tube ± 3–4 mm long, shorter than or subequal to the calyx, the lobes ± 3–4 × 2–2.5 mm, broadly ovate, the margins ± incurved towards the obtuse or rounded apex; *anthers* ± 1–1.5 mm long, shortly exserted on filaments 1–1.5 mm long; *pollen grains* papillate; *style* ± 1–1.5 mm long, extending about half way up the corolla tube; *ovary* ± 0.9 mm long, subtended by 5 small truncate scales ± 0.3 mm long.

Distribution   Alpine and subalpine tracts of the Kosciusko area, N.S.W., and the Bogong High Plains area, Victoria.

Notes and   This species is easily recognized by its conspicuous reddish brown bracts and
Habitat   sepals. It is common in and around the margins of bogs, often forming extensive communities, and in other wet sites, e.g. the margins of fens and short alpine herbfield, along creek banks, and wet stony areas in sod tussock grassland. Populations of the plant have a characteristic reddish purple hue in the autumn, apparently due to extensive production of anthocyanins in the leaves. When growing prostrate over flat rocks, the stems occasionally become quite thick and woody, up to about 1 cm in diameter.

E. microphylla R. Br., Prodr. Flor. Nov. Holl. 550 (1810)
'Coral heath'

Very variable shrub, usually less than 0.5 m high above the treeline, with ascending, decumbent or prostrate stems rooting towards the base (± erect and up to 1 m or more high at lower altitudes); *branchlets* puberulent; *leaves* glabrous, shortly petiolate, rigid, ± 2–4 × 1.8–3 mm, broadly ovate to broadly ovate-cordate, concave above, obtuse, subacute or shortly acuminate and ± pungent at the apex; *petiole* ± 0.5 mm long, articulate on the stem; *flowers* white, solitary and subsessile in the upper axils, forming small leafy spikes in well-grown specimens; *bracts* ovate, chartaceous, pale stramineous; *sepals* ovate, obtuse, 2–3 mm long; *corolla* campanulate, 4–6 mm diam., the tube 1.5–2 mm long, shorter than the calyx, the lobes broadly ovate, blunt, about as long as the tube; *anthers* ± 0.5 mm long, scarcely exserted on very short filaments inserted in the throat of the tube; *style* ± 0.5 mm long; *ovary* ± 0.5 mm long, subtended by 5 small ovate-oblong scales.

See Notes below.

The above description applies to the alpine and subalpine form of this variable species which, in the broad sense, extends along the coast and Great Dividing Range from Queensland to Victoria, and also Tasmania. An important component of *Epacris–Chionohebe* feldmark, also in heaths and bogs. In exposed feldmark sites, *E. microphylla* (and *E. petrophila*) grow very slowly, with rates of stem-diameter increase and stem elongation of as little as 0.2 mm and 10 mm per annum respectively. In such sites, exposure and wind-blast on the windward side of the plant kill the most exposed shoots and branches, whilst on the leeward side where the branches are more protected and wind-blown soil and debris tend to accumulate, the plant slowly grows downwind by the development of roots and shoots from the partly buried branches (this process is called 'layering'). Thus, individual plants tend to migrate slowly in the direction of the prevailing wind, and a given area of ground is alternately covered by an *Epacris* clump and then becomes bare of vegetation. A 'cycle' of development during which given areas are alternately bare and *Epacris*-covered is about 26 years. Other, less hardy species grow within or in the lee of the *Epacris* clumps, and are consequently also being slowly killed and regenerated in cyclical fashion.

Reference: Barrow *et al.* (1968).

Distribution

Notes and Habitat

[252–3]

E. petrophila Hook. f., Flor. Tas. 1: 261 (1857)   'Snow heath'

Small erect shrub to about 40 cm high, or stems prostrate and rooting; *branchlets* puberulent, ± hidden by the leaves; *leaves* sessile, thick, bright green and shining, imbricate, suberect or almost appressed, ± 2.5–3 × 0.8–1.4 mm, elliptic, ovate-oblong or obovate-oblong, slightly concave above, obscurely keeled distally on the underside, subacute, or blunt and slightly upturned at the apex, the margins obscurely serrulate; *flowers* ± 6–7 mm diam., solitary and subsessile in the upper axils, forming small terminal clusters; *bracts* 9–14, pale stramineous becoming brown with age, sometimes faintly tinged with red-brown, the lower pairs small and subdistichous; *sepals* 2.5–3.5 × ± 1.5 mm, ovate, ciliolate, acute or subobtuse; *corolla* white, the tube 1.5–2 mm long, shorter than the calyx, the lobes ovate

[254]

to ovate-triangular, obtuse, 1.5–3 mm long; *anthers* at the throat of the tube, 0.5–0.8 mm long, on short filaments about as long; *pollen grains* smooth; *style* ± 0.5 mm long, about as long as the ovary; *ovary* subtended by five small scales.

Distribution   Alpine and subalpine tracts of the Australian Alps and mountains of Tasmania.

Notes and   A dominant in some areas of *Epacris–Chionohebe* feldmark, usually in somewhat
Habitat   less exposed sites than *E. microphylla* (see Notes on *E. microphylla*). Also found in heaths and bogs.

Pentachondra R. Br.

P. pumila (J. R. & G. Forst.) R. Br., Prodr. Flor. Nov. Holl. 549 (1810)   'Carpet heath'

[255–6]   Small prostrate subshrub 5–10(–15) cm high, usually creeping over low rocks in dense mats or carpets up to 1 m or more in diam.; *stems* long-creeping and rooting, tough and wiry, the branchlets rough with persistent leaf-scars; *leaves* crowded, shortly petiolate, suberect, stiff and shining, 3–6 × 0.5–2.5 mm, oblong or elliptic with rounded or obtuse-callous tips, slightly concave above, glabrous or the young leaves ciliate distally; *petioles* 0.5–1 mm long, articulate at the base; *flowers* white, 5-merous, solitary and sessile near the tips of the branchlets, subtended by 4 or more small imbricate bracts; *sepals* chartaceous, striate, ± 2 mm long, imbricate at the base, with blunt ciliate tips; *corolla tube* cylindrical, ± 4 mm long, hairy inside except at the base, the densely bearded lobes ± 2 mm long; *anthers* ± 0.8 mm long, on short filaments attached near the top of the corolla tube; *ovary* subtended by 5 minute scales; *style* simple, 1–2 mm long; *fruit* drupaceous, crimson, fleshy, globose, 5–8 mm diam. when ripe, crowned by the persistent style and sometimes by the shrivelled corolla; $2n = 28^*$.

Distribution   Alpine and subalpine tracts of the Australian Alps, and mountains of Tasmania and New Zealand (North and South Islands and Stewart Island).

Notes and   Widespread above treeline in heaths, tall alpine herbfields, sod tussock grassland
Habitat   and *Epacris–Chionohebe* feldmark. The green fruit over-winter and ripen early in the following summer; this makes the plant particularly attractive, as the crimson ripe fruits and new season's fresh white flowers form a beautiful contrast on the green carpet of leaves.
Reference: Smith-White (1955)*.

Leucopogon R. Br.

L. montanus (R. Br.) J. H. Willis, Victorian Nat. 73: 56 (1956)   'Snow beard-heath'

[257–8]   Low shrub (10–)30–60 cm high; *stems* tough and wiry, ascending, ± decumbent and rooting towards the base, the branchlets hoary-puberulent; *leaves* glabrous, stiff, shortly petiolate, 3–5(–8) × 1–1.5 mm, narrowly elliptic-obovate, sub-obtuse, green and smooth above, often whitish between the nerves on the under-side, the margins often narrowly translucent and ciliolate distally; *flowers* clustered in short terminal and subterminal spikes, each flower subtended by a small bract and two keeled bracteoles about 1 mm long; *sepals* 5, imbricate, suborbicular, obtuse, ± 1.5 mm long; *corolla* white, the tube 1–1.5 mm long and subequal in length to the sepals, the lobes glabrous or minutely papillate, about as long as the tube; *male flowers* with anthers 0.6–1 mm long; *female flowers* with rudimentary anthers and a short style ± 0.5 mm long; *fruit* drupaceous, subglobular, ± 4 mm diam., red at maturity; $2n = 28, 42$ (as *Lissanthe montana*)*.

Alpine tract of the Australian Alps and high mountains of Tasmania.

Distribution

For a discussion of the very complex breeding system in this species (as *Lissanthe montana*), see Smith-White (1959). It is common in heaths and the small white flowers and red fruit are quite attractive. *L. montanus* is similar in appearance, and apparently closely related, to the more widespread *L. suaveolens* Hook. f. (syn. *L. hookeri* Sond.), which is common in the subalpine tract and at lower altitudes, and some populations near the treeline appear to show characteristics intermediate between the two; the latter species differs mainly in the larger flowers with bearded corolla lobes, the corolla tube ± 2.5–3.5 mm long shortly exceeding the ± 2 mm long sepals and the longer style 1–1.8 mm long.

Notes and Habitat

References: Smith-White (1955*, 1959); Willis (1956*b*).

Gentianaceae

Gentianella Moench

G. diemensis (Griseb.) J. H. Willis, Victorian Nat. 73: 199 (1957)
'Mountain gentian'

Glabrous loosely tufted herb ± 10–25(–30) cm high from a rather weak tap-root; *stems* quadrangular, green or shining reddish purple, erect or ascending from a shortly decumbent base; *basal leaves* subrosulate, 1–5(–7) cm long, narrowly obovate-spathulate to oblanceolate, narrowed to a long or short pseudopetiole; *cauline leaves* opposite, distant along the stem, the uppermost sessile and stem-clasping; *flowers* 5-merous, white or cream with fine bluish or purplish veins, in loose corymbose cymes or occasionally solitary; *calyx* 7–11 mm long, deeply lobed to below the middle, the lobes linear-lanceolate; *corolla* about twice as long as the calyx, broadly campanulate, deeply divided into obovate-cuneate to elliptic lobes, the tube ± 3–4 mm long with 5 nectar pits inside at the base alternating with the stamens; *stamens* epipetalous, the filaments somewhat flat-tened; *anthers* versatile, ± 2 mm long, purplish, rotating to become extrorse after dehiscence; *ovary* shortly stipitate, 1-celled; *stigma* sessile, 2-lobed, persistent

[259–60]

on the ripe capsule; *capsule* ± 2 cm long, narrowly cylindrical, opening in 2 valves at the summit, surrounded by the dried persistent corolla.

Distribution   Mountains of south-eastern Australia from the Barrington Tops, N.S.W., to the Mt Lofty Range, S.A., and mountains of Tasmania.

Notes and   Widespread and common in the alpine and subalpine tracts, especially in sod
Habitat   tussock grassland where its masses of delicate white crocus-like flowers appear rather later in the season (February–March) than most other species. A very variable species over its entire range, requiring further study. Philipson (1972) points out that the generic status of the southern hemisphere gentians is at present uncertain.

Boraginaceae

Myosotis L.

M. australis R. Br., Prodr. Flor. Nov. Holl. 495 (1810)

[261]   Slender tufted erect or ascending perennial herb 15–30 cm high; *stems* simple or branched, covered with spreading hairs except the upper part of the inflorescence with antrorsely appressed hairs, the base of the stem usually with persistent brown remains of old leaves; *leaves* elliptic, oblanceolate or obovate-spathulate, obtuse or rounded at the apex, the upper surface densely hispid with erecto-patent hairs, less conspicuously hairy on the underside, the basal and lower cauline leaves petiolate, the upper leaves sessile and stem-clasping; *petioles* expanded and stem-clasping at the base, those of the basal leaves longer than or subequal to the blade; *inflorescence* a raceme-like cyme, coiled at first, elongating at flowering, ebracteate or with a few leaf-like bracts below; *flowers* white or cream on short pedicels 1–2 mm long, becoming ± secund on recurved pedicels 1–5 mm long in fruit; *calyx* ± 3–4 mm long, enlarging to ± 5–7 mm in fruit, the lobes triangular, ± 2–3 times as long as the tube, densely covered with appressed hairs and sparse erecto-patent hairs, the tube with long spreading hooked hairs; *corolla* rotate, the limb ± 5–8 mm diam., the tube 4–6 mm long, the throat partially closed by 5 yellow notched gland-like scales; *stamens* 5, enclosed in the tube, the filaments inserted about two-thirds up the tube; *style* included,

± 3 mm long, the stigma small; *nutlets* smooth and shining, ovate in outline, ± 2.3 × 1.3–1.5 mm.

See Notes below.

The above description applies to the alpine form of the widespread *M. australis*. It has much larger flowers and nutlets than the forms from lower altitudes and may prove to be a distinct species. It is uncommon above the treeline, found in rocky sites and on moist rock faces, mainly in *Brachycome–Danthonia* tall alpine herbfield.

Labiatae

Prostanthera Labill.

P. cuneata Benth. in DC., Prodr. 12: 560 (1848)

'Alpine mint-bush'

Dark green aromatic spreading shrub 0.5–1.5 m high, the whole plant densely covered with small whitish sessile glands; *branchlets* pubescent with short white spreading hairs; *leaves* coriaceous, opposite, subsessile, 4–8 × 3–5 mm, obovate-cuneate to suborbicular, entire, densely gland-dotted, dark green and shining above, paler on the underside; *flowers* shortly pedicellate, solitary in the upper axils, forming ± leafy racemes; *pedicels* 1.5–2.5 mm long; *calyx* gland-dotted, 5–7 mm long, 2-lipped, the lips rounded, ciliolate, the tube strongly ribbed, 2–3 mm long, subtended by a pair of small narrowly oblanceolate bracteoles about as long as the tube; *corolla* 2-lipped, about 2 cm across, white to pale violet, conspicuously violet-spotted in the throat, puberulent outside and sparsely so inside, the tube ± 7 mm long, the upper lip bilobed, the lower deeply 3-lobed with the middle lobe expanded and deeply cleft or emarginate distally and yellow-blotched proximally near the throat; *stamens* in 2 pairs, the filaments strongly incurved; *anthers* 1–1.5 mm long, the connective on one side produced into a prominent appendage as long as or longer than the cells; *style* arched, slender, ± 9 mm long, minutely forked at the apex; *nutlets* 4, ellipsoid, wrinkled, 2–2.3 × 1–1.5 mm, enclosed within the slightly enlarged persistent calyx at maturity.

Alpine and subalpine tracts of the Australian Alps as far north as the A.C.T., and local near Launceston in Tasmania.

A common shrub in heaths and noteworthy both for the attractive flowers and for the aromatic foliage. Alpine Mint-bush is cultivated in gardens in Canberra and elsewhere.

References: Yeo (1968); Sealy (1950).

Scrophulariaceae

KEY

1 Stamens 4, the anthers hairy, connivent, spurred at the base
Euphrasia (p. 216)
1 Stamens 2, the anthers not as above
2

2 Small much-branched subshrub to ± 10 cm high; corolla funnelform, 5–6-lobed, 1.5–2.5 cm diam.; leaves thick, closely quadrifariously imbricate *Chionohebe densifolia* (p. 219)
2 Slender herbs; corolla small, rotate, 4-lobed; leaves not as above
Veronica (p. 219)

Euphrasia L.

KEY TO THE
SPECIES

The taxa of *Euphrasia* are often very variable and weakly differentiated and some dried specimens are difficult to key out. The unpublished names, designated 'ined.', are used here with the permission of Dr W. R. Barker and will be validated in his forthcoming monograph of the genus in Australia. Dr Barker also provided information on the distribution of the subspecies.

The length of the corolla tube is measured in a straight line from the base to the sinus between the upper and lower lips.

1 Dwarf annual less than 10 cm high; corolla ± 1 cm wide, with conspicuous dark veins, the tube glabrous inside *E. alsa* (p. 217)
1 Perennials, otherwise not as above
2

2 Flowers white (sometimes the buds and young flowers very faintly tinged with pale violet outside); corolla tube 6–9 mm long; style ± 8–10 mm long; wet situations *E. collina* subsp. *glacialis* (ined.) (p.217)
2 Flowers in shades of violet or lilac, or if white then the corolla tube more than 9 mm long and the style more than 10 mm long; drier areas than the preceding
3

3 Corolla tube 9–12 mm long; stems usually more than 15 cm high, ± erect from a decumbent base, the upper cauline internodes twice or more than twice as long as the leaves; widespread in herbfields, grasslands etc.
E. collina subsp. *diversicolor* (ined.) (p. 218)
3 Corolla tube less than 9 mm long; stems usually less than 15 cm high, rather weak and ascending, the upper cauline internodes rarely up to twice as long as the leaves; *Epacris–Chionohebe* feldmark
E. collina subsp. *lapidosa* (ined.) (p. 218)

E. alsa F. Muell., Trans. Phil. Soc. Vict. 1: 107 (1855)
'Dwarf eye-bright'

Small annual 1.5–8 cm high; *stems* erect, simple or with a few branches at or [265]
below the middle, pubescent with short stiff white eglandular hairs and
glandular-septate hairs; *leaves* sessile, up to ± 1 cm long, oblong-cuneate to
obovate-cuneate in outline, trifid or pinnatifid with 5 obtuse lobes, scaberulous
with short eglandular hairs and sparser glandular-septate hairs; *bracts* similar to
the upper cauline leaves; *flowers* subsessile or shortly pedicellate in short dense
racemes; *calyx* darkly veined, 3.5–7 mm long, enlarging in fruit, puberulent with
glandular and eglandular hairs, the 4 lobes narrowly triangular, obtuse, shorter
than the tube; *corolla* 2-lipped, ± 8–10 mm across, white to pale mauve with
conspicuous deep violet longitudinal veins, pubescent outside, the tube 4–7 mm
long and glabrous inside, the upper lip bilobed, pubescent near the sinus, the
lower lip deeply trilobed with spreading emarginate lobes; *anthers* 4, connivent,
bearded, spurred at the base; *style* filiform, ± 5 mm long, puberulent distally;
capsule elliptic, 4–7 × 2.5–3.5 mm, compressed above, subequal to or shorter
than the calyx, ± pubescent and pilose on the margins in the upper half.

Alpine and subalpine tracts of the Kosciusko area, N.S.W.   *Distribution*

Common in *Epacris–Chionohebe* feldmark, also in bare areas in sod tussock grass-   *Notes and*
land and tall alpine herbfield. Rare in the subalpine tract.   *Habitat*

E. collina R. Br. subsp. *glacialis* (Wettst.) Barker (ined.)

Perennial herb ± 5–15 cm high, usually with short non-flowering shoots at [266–7]
the base; *flowering stems* simple, ± erect or ascending from a decumbent base,
usually dark and shining, almost glabrous or with inconspicuous lines of hairs
towards the base, hirsute distally with septate and glandular-septate hairs; *leaves*
sessile, mostly opposite, those on the non-flowering stems and towards the base
of the flowering stems crowded, subglabrous or sparsely puberulent, the upper
cauline leaves becoming larger and more distant up the stem, 0.5–1 cm long,
broadly cuneate to obovate-oblong in outline, subdigitately 3–5-toothed, ±
puberulent with glandular and non-glandular hairs; *flowers* solitary in the axils
of leafy bracts, in short dense corymbose clusters; *calyx* 4-toothed, darkly veined,
± 6–8 mm long, densely pubescent with septate and glandular-septate hairs
except towards the base; *corolla* 2-lipped, the upper lip hooded with 2 reflexed
lobes, the lower lip with 3 spreading lobes, white, yellow in the throat (sometimes
the buds and young flowers faintly tinged with pale violet on the outside),
shortly hairy on the outside and inside the upper lip, the tube 6–9 mm long,
hairy inside towards the base; *stamens* 4, didynamous, the anthers connivent,
hairy on the back and along the lines of dehiscence, the cells spurred at the
base; *style* slender, ± 8–10 mm long; *capsule* loculicidal, glabrous or sparsely hairy
distally, ± 6–7 × 3.5 mm, elliptic or elliptic-oblong, rounded or subtruncate
and compressed at the apex.

Alpine tract of the Kosciusko area, N.S.W.   *Distribution*

Common in wet areas and depressions in sod tussock grassland and short alpine   *Notes and*
herbfield, and wet flushes in tall alpine herbfield. This attractive little subspecies   *Habitat*

is usually in mass flower towards the end of December and in the first half of January, although small patches can be found in flower until March in especially wet sites.

The species of *Euphrasia* are hemi-parasites, their roots becoming attached to those of adjacent plants by means of small haustoria. The undersides of their leaves have shallow depressions which are densely lined with microscopic glands.

E. collina R. Br. subsp. *diversicolor* Barker (ined.)

[268–9] Generally more robust and often more densely pubescent than the preceding, the stems up to 30 cm high, often more stiffly erect from the decumbent bases, and otherwise differing as follows: upper cauline leaves usually larger, ± 1–1.5 cm long, oblong to broadly oblong-cuneate in outline, usually coarsely 5–9-toothed; *flowers* larger, the *calyx* 7–12 mm long and less conspicuously veined; *corolla* in shades of violet, lilac or white, the tube 9–12 mm long; *style* ± 12–15 mm long; *capsule* larger, ± 7–11 mm long.

Distribution Alpine and subalpine tracts of the Kosciusko area, N.S.W. (with related variant in the Cobberas Mountains, Victoria).

Notes and Habitat This subspecies is the most common above the treeline and is easily distinguished *en masse* from the preceding by the larger, variably coloured flowers. It is usually found in drier sites in sod tussock grassland, tall alpine herbfields and sometimes in heaths, and flowering usually begins about mid January, when the main flowering of the other alpine subspecies is on the wane.

E. collina R. Br. subsp. *lapidosa* Barker (ined.)

[270–1] This subspecies is usually smaller than the subsp. *diversicolor*, and is frequently quite dwarfed in exposed parts of the feldmark; the stems are ± 3–15 cm high, rather weak, often numerous and decumbent or ascending from a decumbent base; it differs otherwise from the subsp. *diversicolor* mainly in the more crowded leaves with shorter upper cauline internodes, and in the generally smaller flowers with calyx ± 7–10 mm long and corolla tube ± 7.5–8.5 mm long.

Distribution Endemic to the alpine tract of the Kosciusko area, N.S.W.

Notes and Habitat This subspecies is restricted to the *Epacris–Chionohebe* feldmark. The main flowering of this subspecies is usually completed before the subsp. *diversicolor* comes into full bloom. The pale violet to purplish pink flowers of this plant are sometimes remarkably similar in colour to those of *Chionohebe* with which it grows, and Yeo (1968) instances a similar situation in *Euphrasia micrantha* of Europe, the flowers of which appear to mimic those of Heather (*Calluna vulgaris*). He also draws attention to the general similarity of floral characters between *Euphrasia* spp. and *Prostanthera cuneata*, and these observations suggest that a detailed study of the pollinators of these species in the Kosciusko area would be rewarding.

Reference: Yeo (1968).

Chionohebe Briggs & Ehrend.

C. densifolia (F. Muell.) Briggs & Ehrend., Contrib. Herb. Aust.
No. 25: 2 (1976)

Low trailing or creeping subshrub forming tufts or patches ± (5–)10–30 cm   [272–3]
diam.; *stems* prostrate or decumbent, freely rooting, the short ascending
branchlets pubescent and up to about 10 cm high; *leaves* opposite, sessile, shortly
connate at the base, thick and coriaceous, closely quadrifariously imbricate,
entire, ± 4–7 × 2–3 mm obovate-oblong to obovate-spathulate, blunt, concave
above, obscurely keeled on the underside, fringed with simple and/or glandular-
septate hairs proximally, the margins thickened and cartilaginous distally; *flowers*
± 1.5–2.5 cm diam., solitary and subsessile in the upper axils, subtended by
two leaf-like bracteoles; *calyx* 5–7 mm long, divided to the middle or below
into 5(–6) oblong, obtuse lobes which are ciliate and glandular-pubescent proximally, the tube ± rugose; *corolla* broadly funnelform, mauve in bud, pale violet
to purplish pink, or white, divided to the middle or below into 5–6 broadly
obovate-spathulate to suborbicular lobes, the tube 4–5 mm long; *stamens* 2, the
filaments 2–4 mm long, inserted in the throat; *anthers* ± 1.5 mm long, the cells
divergent; *style* slender, 4–6 mm long, the stigma small and capitate; *capsule* narrowly obcordate, ± 3.5–4 ×3 mm, glabrous, or sometimes pubescent towards
the apex when young.

Alpine tract of the Mt Kosciusko area, N.S.W., and the South Island of New   *Distribution*
Zealand (higher mountains of the Lake District, central and north Otago).

In the Kosciusko area this species is restricted to the high feldmarks of the Main   *Notes and*
Range from Mt Lee to Mt Twynam and to similar sites on the Etheridge Range.   *Habitat*
It was until recently thought to be endemic to this area. However, Briggs and
Ehrendorfer (1976) have indicated that the New Zealand species formerly
known as *Pygmea tetragona* (Hook. f.) M. B. Ashwin is synonymous with the
Kosciusko plant. *C. densifolia* is the only species known from the mainland, and
it has recently been discovered (Ratkowsky 1974) that Tasmania also shares
a species of this small genus with New Zealand. At Kosciusko, this species grows
in close association with *Euphrasia collina* subsp. *lapidosa* (ined.) and the flower
colour of the two plants is sometimes very similar (see Notes under the latter
species). It is early-flowering, mainly from mid November to mid January.

References: Allan (1961); Briggs and Ehrendorfer (1976); Mark and Adams
(1973); Ratkowsky (1974).

Veronica L.

V. serpyllifolia L., Sp. Pl. 1: 12 (1753)   'Thyme speedwell'

[274]   Slender perennial herb to 25 cm high; *stems* erect or ascending, creeping and rooting at the base, puberulent with septate and glandular-septate hairs; *leaves* opposite (rarely in whorls of 3), 0.5–1.2 cm long, the lower petiolate, the upper subsessile, glabrous or sparsely puberulent, ovate, oval or oblong, entire or weakly crenulate; *racemes* lax, ± glandular-puberulent, the upper bracts narrow-oblong, the lower similar to the leaves; *flowers* zygomorphic; *calyx* 4-lobed, the lobes oblong, 1–2 mm long in flower, enlarging in fruit; *corolla* rotate, pale mauve (occasionally almost white) to violet, with darker veins, ± 5–6 mm diam., with a short tube about 1 mm long, 4-lobed, the lower lobe smaller than the rest; *stamens* 2; *style* slender, subpersistent, elongating to about 3 mm in fruit; *capsule* broadly obcordate, about 3 × 4–5 mm, minutely glandular-pubescent above.

Distribution   Mountainous areas of eastern N.S.W. and Victoria and widespread in Tasmania; temperate regions throughout the world.

Notes and   Rare above the treeline, in sod tussock grassland and wet seepage areas, e.g.
Habitat   near the base of the crags at the western end of Blue Lake, and margins of swamps and streams.

Plantaginaceae

Plantago L.

3 Leaves (3–)5(–9)-nerved, the nerves mostly equally conspicuous on the upper surface and all extending to the distal half of the lamina; fruiting spikes (2–)4–8(–12) cm long .... *P. euryphylla* (p. 222)

P. glacialis Briggs, Carolin & Pulley, Contrib. N.S.W. Natl Herb. 4: 395 (1973)

Small perennial herb usually only a few cm high from a long, much-branched [275–6] erect or ascending rootstock; *stem* densely covered with pale ferruginous hairs, and with the remains of old leaves towards the base; *leaves* thick and shining, 1-nerved, dotted with minute hydathodes, glabrous or sometimes the margins or upper surface with long septate hairs, (1–)2–4(–7) cm × 1–5 mm, incl. the broad petiole, narrowly elliptic to narrowly oblanceolate, entire or with distant obscure teeth; *scapes* not or only shortly exceeding the leaves, glabrous or occasionally with sparse long septate hairs; *spike* small, capitate, ± 2.5–6.5 mm long with (1–)2–3(–5) flowers; *bracts* glabrous, concave, dark brown, 1.5–3 × 1–2 mm, conspicuously keeled; *sepals* glabrous, blunt, elliptic, 2–2.2 × 1–2.2 mm; *corolla* scarious, the lobes ovate, 0.7–1.2 mm long; *stamens* 4, the anthers exserted on filaments 4–7 mm long; *anthers* 1 mm long; *ovary* 2-celled with 2 ovules in each cell; *style* 5–6 mm long; *capsule* ellipsoid, 1.5–2 mm diam.; *seeds* reddish brown, elliptic, 1.2–1.8 × 0.8–0.9 mm; $2n = 12^\star$.

Alpine tract of the Kosciusko area N.S.W., and the Bogong High Plains, Victoria. *Distribution*

This species is a characteristic component of short alpine herbfield, usually form- *Notes and* ing extensive, closely appressed carpets or mats of star-shaped plants below snow *Habitat* patches. In good seasons, the rosettes of individual plants are often so closely crowded together that they form a beautifully regular hexagonal pattern. We suggest that an appropriate common name for this species would be Small Star Plantain.

References: Briggs (1973)*; Briggs *et al.* (1977).

P. muelleri Pilger, Pflanzenreich IV 269 (Heft 102): 118 (1937)
'Star plantain'

Perennial herb from a short erect rootstock with copious long fleshy adventitious [277–8] roots; *stem* densely covered with long ferruginous hairs; *leaves* densely rosulate, stiff, thick and glabrous, fleshy, dark green and shining, dotted with hydathodes, 3-veined, the midvein prominent on the underside, the lateral veins relatively inconspicuous, (1.5–)3–9(–11.5) × (0.5–)1–2.5 cm incl. the broad petiole, elliptic-lanceolate to ovate-lanceolate, the margin entire or with irregular distant teeth; *scapes* usually fairly numerous, antrorsely appressed-pilose, short and almost hidden among the leaves at anthesis, elongating to ± 5–20 cm in fruit; *spike* short, subcapitate or occasionally up to 1 cm long in fruit, 2–13-flowered; *bracts* concave, glabrous, 2.4–3.5 mm long, ovate, obtusely keeled; *sepals* glabrous, acute or subacute, elliptic, 2.5–3.5 mm long; *corolla* scarious, the lobes 1.4–2 mm long, acute; *stamens* 4, the anthers exserted on filaments to 6 mm long; *ovary* 2-celled with 2 ovules in each cell; *style* 5–10 mm long; *capsule* ellipsoid, 2–2.5 mm diam.; *seeds* yellowish brown, elliptic, ± 2 × 1–1.2 mm; $2n = 36^*$.

Distribution   Alpine and subalpine tracts of the Kosciusko area, N.S.W., and Mt Gingera, A.C.T.

Notes and Habitat   Common in wet areas in short alpine herbfield and on the margins of streams and bogs. This species is readily distinguished by the broad, thick, glabrous, stiff glossy leaves and the very short scapes at flowering time. It appears to be closely related to *P. glacialis* but is larger in all its parts.

References: Briggs (1973)*; Briggs *et al.* (1977).

P. alpestris Briggs, Carolin & Pulley, Contrib. N.S.W. Natl Herb. 4: 395 (1973)

[279]   Scapigerous perennial herb from a short erect rootstock, the stem densely covered with long pale ferruginous hairs; *leaves* 3(–5)-nerved, the midnerve distinct, the lateral nerves less conspicuous, (2–)6–8 × (0.7–)1–2(–2.5) cm. incl. the broad petiole, entire, oblanceolate-elliptic to obovate-elliptic, ± pubescent and often glabrescent above; *scapes* pubescent, ± (2–)8–16 cm long; *spike* ovoid to cylindrical, (0.5–)1–4(–6) cm long in the fruiting stage; *bracts* concave, ovate, obtuse, glabrous or with a few long hairs at the apex, 2–2.5 mm long, obtusely keeled; *sepals* subequal, 1.5–1.8 mm long, elliptic, obtuse or rounded at the apex; *corolla* scarious, the lobes ovate, acute, ± 1 mm long; *stamens* 4, the anthers exserted on filaments up to 5 mm long; *ovary* 2-celled with 2 ovules in each cell; *style* 3.5–4 mm long; *capsule* ovoid, 2.3–2.5 mm diam.; *seeds* ± 1.5 × 0.8 mm, elliptic, dark reddish brown at maturity; $2n = 12^*$.

Distribution   Alpine and subalpine tracts of the Kosciusko area, N.S.W., as far north as Mt Jagungal, and Mt Baw Baw and Mt Buffalo in Victoria.

Notes and Habitat   Common in moist sites in several communities including sod tussock grassland and tall and short alpine herbfields, and along the banks of streams.

References: Briggs (1973)*; Briggs *et al.* (1977).

P. euryphylla Briggs, Carolin & Pulley, Contrib. N.S.W. Natl Herb. 4: 396 (1973)

[280]   Scapigerous perennial herb from a short erect rootstock with fleshy adventitious roots, the stem densely covered with ferruginous hairs; *leaves* (3–)5(–9)-nerved, 3–10.5 × 1.3–4 cm, incl. the broad petiole, ovate, elliptic or obovate, ± densely pubescent or hirsute, the margins usually with rather obscure irregular distant teeth; *scapes* antrorsely appressed-pubescent, (6–)10–20(–30) cm long; *spike* narrowly cylindrical, (2–)4–8(–12) cm long in the fruiting stage; *bracts* glabrous except usually pubescent distally, concave, 2.5–4 mm long, keeled, ovate-elliptic, obtuse, ciliate on the margins; *sepals* glabrous, elliptic, obtuse, 1.5–2.5 mm long;

corolla scarious, the tube subequal in length to the sepals, the lobes broadly ovate to orbicular, 0.5–1 mm long; *stamens* 4, the anthers exserted on filaments to 3 mm long; *ovary* 2-celled, with 2 ovules in each cell; *style* 1.5–3 mm long; *capsule* broadly ovoid, 2.5–3.5 × 1.5–2 mm; *seeds* 1.2–1.8 × 0.7–1 mm, elliptic, greenish brown at maturity; $2n = 12^*$.

Alpine and subalpine tracts of the Australian Alps and associated ranges, as far north as the Brindabella Range, A.C.T.   *Distribution*

This species occurs in rather drier habitats than the other alpine species of *Plantago* and may be distinguished from them by the usually broader and hairier, more strongly veined leaves. It is locally common in tall alpine herbfield and drier areas in sod tussock grassland. It may be confused with the subalpine *P. antarctica* Decne; however, the latter differs in having a long persistent tap-root, the root-stock of *P. euryphylla* being very short and truncate, with copious long fleshy adventitious roots.   *Notes and Habitat*

References: Briggs (1973)*; Briggs *et al.* (1977).

Rubiaceae

Asperula L.

A. gunnii Hook. f., Hook. Lond. J. Bot. 6: 463 (1847)

'Mountain woodruff'

[281]   Dioecious perennial herb to 15 cm high from slender horizontal or ascending rhizomes 1–1.5 mm diam.; *stems* decumbent or ascending, 4-angled, lax, much-branched, clustered, minutely retrorsely scabrid on the angles; *leaves* (and leaf-like stipules) usually in whorls of (4–)6, ± 5–10 × 1.5–2.5 mm, glabrous or the slightly recurved margin sometimes ciliolate, sessile or shortly petiolate, obovate, obovate-oblong or oblanceolate, acute or subacute; *flowers* white or creamy, with rudimentary calyx, (3–)4(–5)-merous, sweet-scented, in small clusters in the upper axils; *male flowers: corolla tube* 1–1.5 mm long, the lobes about as long; *stamens* exserted, attached near the summit of the tube and alternate with the lobes; *female flowers: corolla tube* ± 0.5–1 mm long, the lobes ± 1 mm long; *style* exserted, 2-branched towards the summit, with conspicuous capitate stigmas; *fruit* of two glabrous connate globose mericarps, each 1.5–2 mm diam.

Distribution   From the Blue Mountains southwards, on the higher mountains of eastern N.S.W., the A.C.T., eastern Victoria and Tasmania.

Notes and   Fairly common in tall alpine herbfields, sod tussock grassland and heaths, some-
Habitat   times forming patches up to 20 cm or more in diam. Herbarium specimens of this and the following species often become almost black when dried.
Reference: Shaw and Turrill (1928).

A. pusilla Hook. f., Hook. Lond. J. Bot. 6: 464 (1847)

'Alpine woodruff'

[282]   Superficially similar to some small-leaved plants of the preceding, but the stems and leaves more crowded, the whole plant hispid-pubescent, and differing otherwise mainly as follows: *leaves* (and leaf-like stipules) mostly narrower, usually in whorls of 6(–8), ± 2–8 × 0.9–1.5 mm, linear-oblong to oblanceolate, obtuse to subacute or rounded at the apex, densely hispid-pubescent on the upper surface, ± ciliate on the margins and on the midrib below.

Distribution   Alpine and subalpine tracts of the Australian Alps and associated ranges as far north as the A.C.T., and mountains of Tasmania.

Notes etc.   Tall alpine herbfield, apparently less common than *A. gunnii* above the treeline.
Reference: Shaw and Turrill (1928).

Coprosma J. R. & G. Forst.

C. sp.

[283–5]   Prostrate shrub forming patches up to 1 m or more in diam.; *stems* glabrous, creeping and rooting, densely matted or trailing; *leaves* glabrous, thick, shortly petiolate, obovate to elliptic, ± 3–6 × 2–2.3 mm incl. the ± 0.5 mm petiole,

slightly concave and shining above; *stipules* interpetiolar, broadly triangular, ±
ciliate, the terminal denticle ± 0.3 mm long, lateral denticles 0–3 on each side;
flowers usually 4-merous, pale yellow, solitary, sessile and terminal on short
branches, subtended by 2 small bracts; *staminate flowers*: *calyx* 1.5–2 mm long,
the tube ± 1 mm long, usually with 2 short obtuse lobes 0.3–0.5 mm long and
2 longer obtuse or acute lobes 0.5–1 mm long; *corolla* funnelform, the tube ±
3–4 mm long, the lobes 2–2.5 mm long; *stamens* long-exserted, the filaments
± 8–10 mm long, slightly flattened towards the base, the anthers 2.5–3.2 mm
long, incl. the short papillate appendage ± 0.2 mm long; *styles* rudimentary,
or sometimes short styles up to ± as long as the corolla present; *female flowers*:
calyx ± 2–2.5 mm long, usually with 2 long and 2 short lobes; *corolla* tube 1–1.5
mm long, the lobes spreading, longer than the tube, 2.5–3.5 mm long; *stamens*
reduced and sterile, enclosed within the tube or shortly exserted; *style branches*
usually 2, thick and densely papillate, ± 7.5–10 mm long; *drupe* bright orange-red,
ovoid to globular, ± 5–7 mm diam.; *pyrenes* 2, ovate in outline, plano-convex,
3–3.4 × ± 2.2 mm, yellowish.

Possibly restricted to the Kosciusko area, N.S.W. *Distribution*

The *Coprosma pumila* complex needs revision. The Kosciusko plants, which have *Notes and*
until now been referred to *C. pumila*, differ markedly from other specimens *Habitat*
of the complex I have seen in the more robust habit, the shorter corolla tube,
the usually 2 style branches and pyrenes, etc.; other anomalies within the complex
have been indicated by Oliver (1935) and Moore (1964). A characteristic dominant in *Coprosma–Colobanthus* feldmark, where individual plants form spreading
bright green mats on rocky snow-patch areas. In such a difficult environment
the *Coprosma* plants are probably very slow-growing, but once established may
live to a considerable age. This must be regarded as a rare species, the number
of mature plants in the feldmarks probably not greatly exceeding 1000 (Wimbush
and Costin 1973).

References: Moore (1964); Oliver (1935); Wimbush and Costin (1973).

Nertera Banks & Soland. ex Gaertn.

N. depressa Banks & Soland. ex Gaertn., Fruct. et Semin. Plant. 1: 124, t. 26 (1788)

Glabrous prostrate perennial herb, often ± hidden among mosses or forming [286]
small patches in wet sand or gravel; *stems* up to 1 mm diam., much-branched,
creeping and rooting; *leaves* broadly ovate-triangular to suborbicular, blunt or
obtuse, the blade ± 1.5–2 × 1–2 mm, ± abruptly narrowed to the flattened
0.5–2-mm-long petiole; *stipules* interpetiolar, small and shallow with a prominent central denticle, fused with the petiole base and forming persistent sheaths
after the leaves have fallen; *flowers* (not seen by me) minute, terminal, sessile
and solitary; *drupe* succulent, bright orange-red, ovoid to globular, up to 5 mm
diam., with two pale, oval, plano-convex pyrenes ± 2 × 1.5 mm.

See Notes below. *Distribution*

Notes and Habitat As Willis (1973) points out, this genus is badly in need of revision, and until this is done, the above name can only be regarded as provisional. This species is found in short alpine herbfield and in wet sites on the margins of streams, fens and bogs. It is a very inconspicuous little plant and is usually only noticed when the attractive bright orange fruit appear in November.

Reference: Willis (1973).

Campanulaceae

Wahlenbergia Schrad. ex Roth

W. ceracea Lothian, Victorian Nat. 72: 166 (1956)

'Waxy bluebell'

[287] Perennial herb 15–30(–60) cm high from slender branched rhizomes; *stems* slender, at first decumbent, finally ascending or erect, usually simple, glabrous above, sparsely hairy towards the base; *leaves* mostly alternate, (1–)2–3 cm long, glabrous or almost so, oblanceolate to spathulate, undulate or almost flat, irregularly crenate-dentate without raised calli on the teeth; *flowers* ± 3–4.5 cm diam., pale blue to pale violet, nodding in bud; *sepals* narrowly triangular, 4–6 mm long; *corolla* campanulate, the tube 6–9 mm long, whitish and pubescent inside, the lobes more than twice as long as the tube; *stamens* 5, the filaments abruptly expanded and ciliate at the base; *style* sparsely glandular above, exceeding the corolla tube; *stigmas* (2–)3; *capsule* broadly obovoid, ± 6–9 × 5 mm, crowned by the persistent erect sepals.

Distribution Alpine, subalpine and upper montane tracts of the Australian Alps and associated ranges as far north as the Brindabella Range, A.C.T., the Barrington Tops, N.S.W., and mountains of Tasmania.

Notes and Habitat Widespread in the alpine tract especially in sod tussock grassland and wet areas near stream banks and the margins of bogs and fens. This bluebell makes a spectacular showing with its sky-blue flowers late in the season (Feb.-March), when most other species have finished flowering. Dried specimens from the Kosciusko alpine tract are strongly coumarin-scented.

W. gloriosa Lothian, Proc. Linn. Soc. N.S.W. 71: 224 (1947)
'Royal bluebell'

Differs from *W. ceracea* mainly as follows: *stems* decumbent or ascending and usually more hirsute towards the base; *leaves* mostly opposite, usually more crowded, obovate, obovate-oblong or spathulate, subglabrous or variously stiffly hirsute, especially on the underside, the margins conspicuously crisped-undulate and coarsely toothed, the teeth usually with a small raised callus tip; *flowers* deep purplish violet; *stigmas* 2(–3); *capsule* ± 10–15 × 5 mm. [288]

Mainly in subalpine woodland in the Australian Alps and adjacent ranges as far north as the Brindabella Range, A.C.T. *Distribution*

Although this spectacular species is found in abundance in the subalpine tract, it is not common above the treeline; it has been collected from Sentinel Peak and around Blue Lake and is usually found growing among rocks. *Notes and Habitat*

Lobeliaceae

Pratia Gaudich.

P. surrepens (Hook. f.) F. E. Wimm., Pflanzenreich IV 276 b (Heft 106): 108 (1943) 'Mud pratia'

Small prostrate perennial herb with milky juice; *stems* long-creeping and freely rooting, ± 1.5 mm diam., glabrous, or sparsely puberulent at the tips; *leaves* alternate, in one plane, pressed close to the ground, the blade obovate-spathulate to suborbicular, ± 5–10 × 3–6 mm, usually entire, glabrous or sometimes ciliolate towards the base, rarely puberulent on the upper surface, narrowed to a short sheathing flattened petiole; *flowers* irregular, solitary in the axils, on pedicels shorter than or subequal to the leaves; *calyx* glabrous or puberulent, the lobes 5, triangular, 1.5–2.5 mm long; *corolla* 5-lobed, white or tinged with pale violet, 10–13 mm diam., glabrous or puberulent on the outside, the tube subequal to or exceeding the calyx lobes and split to the base on the upper side; *stamens* attached near the base of the corolla tube; *anthers* purplish, ± 1.5 mm long, cohering in an oblique tube around the style, the two lower anthers tipped with minute bristles; *ovary* inferior; *fruit* a fleshy, indehiscent glabrous or puberulent berry. [289]

Alpine and subalpine tracts of the Australian Alps, and mountains of Tasmania. *Distribution*

Locally common on peat in fen and bogs, and on mud in depressions in sod tussock grassland. *Notes and Habitat*

Goodeniaceae

Goodenia Sm.

G. hederacea Sm. var. *alpestris* Krause, Pflanzenreich IV 277 (Heft 54): 56 (1912)

[290] Prostrate perennial herb; *stems* leafy, procumbent or long-creeping and rooting at the nodes, white-tomentose towards the tips, the older stems puberulent or glabrescent; *leaves* erect, broadly ovate to broadly elliptic-oblong or suborbicular in outline, blunt or obtuse, 1–3 × 1–2.5 cm, coarsely crenate-dentate or serrate, sometimes irregularly lobed at the base, glabrescent above, white appressed-tomentose on the underside, ± abruptly narrowed to the slender petioles which are subequal to or longer than the blades; *flowers* yellow, solitary in the axils, the pedicels slender, longer than the leaves, with two small subulate bracteoles at or below the middle; *calyx* 6–8 mm long, tomentose except the nerves glabrescent, the lobes subulate, subequal to or shorter than the tube; *corolla* 1–1.3 cm long, the short tube slit to the base on the upper side, the lobes hairy on the outside, the broad, yellow glabrous wings deeply notched at the apex; *stamens* 5, free, ± 4 mm long; *style* ± 4 mm long, the broad cup-shaped indusium pubescent towards the base and densely ciliate distally; *capsule* 2-valved, obovoid, 6–7 mm long, crowned by the persistent sepals; *seeds* yellowish, 2.5–2.8 × 1.5–1.7 mm, elliptic, flattened with minutely prickly surface and slightly thickened margins.

Distribution   (of the variety): Alpine and subalpine tracts of the Australian Alps and associated ranges as far north as the A.C.T.

Notes and Habitat   This plant is common in the subalpine tract but marginal above the treeline in tall alpine herbfield.

Stylidiaceae

Stylidium Swartz ex Willd.

S. graminifolium Swartz ex Willd., Sp. Pl. 4: 146 (1805)
'Grass trigger-plant'

[291–3] Tufted perennial herb 20–40 cm high from a short ± branched rootstock; *scapes* usually reddish, minutely glandular-hairy, especially above; *leaves* radical, densely tufted, rather stiff, glabrous, linear and grass-like, ± 5–10 × 0.2 cm, sharply pointed, the margins distantly serrulate; *flowers* irregular, shortly ped-icellate in a spike-like raceme, each flower subtended by two minute linear-subulate bracteoles and a broadly ovate bract 2–3 mm long; *calyx* minutely glandular-hairy, the lobes ± connate into 2 broad lips; *corolla* deep pink to pur-plish red, deeply divided into 4 large subequal laterally paired lobes and a small, inconspicuous reflexed lobe (the labellum) which has 2 minute linear appendages near the base, the throat of the tube with ± 6 conspicuous linear appendages; *stamens* 2, adnate to the style, forming a long flattened geniculate irritable column (the gynostemium or 'trigger'); *anthers* with divergent cells, fringed with hair-like papillae on the back; *stigma* sessile between the anthers; *ovary* inferior, with a large nectar-secreting gland at the summit; *capsule* ovoid-cylindrical, 2-valved, 5–10 mm long, surrounded by the persistent calyx.

See Notes below.

The above description applies only to the alpine and subalpine form of this wide-spread and variable species. The deep pink to purplish red flowers form a beautiful display late in the season (Feb-Mar.). It is common in a variety of communities including tall alpine herbfields, sod tussock grassland, heaths, fen and bogs. The ingenious trigger action of the column ensures cross fertilization; the column at rest is bent backwards over the labellum towards the calyx tube, and when the base of the column is touched by an insect in search of nectar, the column abruptly springs forward and showers the insect with pollen; at a later stage the receptive stigma receives pollen from another plant. The column can be activated artificially by probing at the base with a pin or the tip of a stiff blade of grass.

Distribution
Notes and
Habitat

Reference: Erickson (1958).

Compositae

1	At least the central florets tubular; plants without milky juice	2	KEY
1	Florets all ligulate; plants with milky juice	19	
2	Pappus conspicuous, more than half as long as achene	3	
2	Pappus short, less than one-third the length of the achene, or absent	16	
3	Capitula radiate	4	
3	Capitula non-radiate	8	
4	Shrubs	*Olearia* (p. 365)	
4	Herbs	5	
5	Phyllaries with broad scarious laminae	*Podolepis robusta* (p. 374)	
5	Phyllaries herbaceous	6	
6	Ligules of ray florets yellow	*Senecio* (in part) (p. 378)	
6	Ligules of ray florets not yellow	7	

7	Leaves linear, silvery	*Celmisia* (p. 364)
7	Leaves not as above	*Erigeron* (p. 363)

8	Inner phyllaries with conspicuous white or yellow petaloid tips	9
8	Inner phyllaries without coloured petaloid tips	12

9	Shrubs	*Helichrysum* (in part) (p. 371)
9	Herbs or low mat-forming subshrub	10

10	Petaloid tips of inner phyllaries yellow	*Helichrysum scorpioides* (p. 372)
10	Petaloid tips of inner phyllaries white	11

11 Low mat-forming subshrub up to ± 10 cm high; capitula less than 1 cm
diam. *Ewartia nubigena* (p. 369)
11 Not as above; capitula more than 1 cm diam. *Helipterum* (p. 370)

12 Capitula numerous in a dense globular or hemispherical compound head
Craspedia (p. 375)
12 Capitula not in a compound head 13

13 Phyllaries herbaceous, mostly in one row except for a few small bractlets
at the base *Senecio gunnii* (p. 378)
13 Phyllaries in more than 1 row 14

14 Phyllaries conspicuously long-ciliate distally *Leptorhynchos* (p. 374)
14 Phyllaries not long-ciliate distally (note that the innermost phyllaries of
Gnaphalium umbricola are deeply laciniate) 15

15 Leaf blades glabrous; dwarf mat-forming plant 1–2 cm high
Parantennaria uniceps (p. 370)
15 Leaf blades hairy, at least on the underside *Gnaphalium* (p. 366)

16	Capitula radiate	17
16	Capitula non–radiate	18

17 Achenes beaked; pappus absent *Lagenifera stipitata* (p. 231)
17 Achenes not beaked; pappus present but sometimes small and inconspicuous *Brachycome* (p. 231)

18	Leaves entire	*Abrotanella nivigena* (p. 377)
18	Leaves pinnatifid	*Cotula alpina* (p. 377)

19 Pappus of narrow flat scales which taper above to barbellate bristles
Microseris lanceolata (p. 380)
19 Pappus of plumose or barbellate bristles 20

20	Pappus bristles (except the outermost) plumose	21
20	Pappus bristles barbellate	22

21 Receptacle scales present; scapigerous herbs without bifurcate-hooked bristles; common in disturbed sites and Soil Conservation reclamation areas
Cat's-ear *Hypochoeris radicata*
21 Receptacle without scales; herbs with leafy stems, hispid with bifurcate-
hooked bristles; rare near roadside Hawkweed Picris *Picris hieracioides*

22 Scapigerous herb; capitula solitary and terminal on hollow scapes; achenes beaked, muricate below the beak; locally common in disturbed sites and Soil Conservation reclamation areas   Dandelion *Taraxacum officinale* sp. agg.

22 Herb with leafy stems; inflorescence branched; achenes not beaked or muricate; locally common on Soil Conservation reclamation areas
                                    Smooth Hawk's-beard *Crepis capillaris*

Lagenifera Cass.

L. stipitata (Labill.) Druce, Rep. Bot. (Soc.) Exch. Cl. Manchr 1916: 630 (1917)

Stoloniferous perennial herb 4–12(–17) cm high from extensive slender long-creeping branched rhizomes about 1–1.5 mm diam.; *scapes* hirsute, usually with a few small linear bracts; *leaves* mostly basal, petiolate, obovate-spathulate, blunt or ± obtuse, densely hirsute on both sides, variably crenate-serrate to sinuate-dentate; *petiole* subequal to or shorter than the blade; *capitula* solitary, scented, the involucre ± 5 × 5–8 mm; *phyllaries* numerous in 3–4 rows, herbaceous, linear-lanceolate, 3.5–4.5 mm long, with a thin longitudinal median strip and thickened sides which are pale when dry, the apex thin, acute or acuminate with sparsely ciliolate margins; *ray florets* ligulate, numerous in 3–4 rows, the tube glandular, the ligules white or mauve and 2–4 mm long; *disc florets* with glandular tubes, functionally staminate; *achenes* compressed, narrowly obovate to oblanceolate, slightly asymmetrical, the margins minutely glandular, 2.2–3 × 0.8–1 mm excluding the curved glandular beak which is 0.5–1 mm long with a thickened pale collar-like apex; *pappus* absent.   [294]

South-eastern Australia including Tasmania, Papua, and New Zealand (North Island where perhaps adventive, see *).   *Distribution*

A widespread but rather inconspicuous minor herb forming patches in sod-tussock grassland and tall alpine herbfields, usually in the shelter of large granite outcrops.   *Notes and Habitat*

References: Cabrera (1966); Davis (1950); Drury (1974)*.

Brachycome Cass.

1	Fruit with conspicuous broad flattened wings	2	KEY TO THE
1	Fruit narrowly margined, not conspicuously winged	4	SPECIES

2 Leaves broad, spathulate to obovate, blunt, crenate or crenate-serrate
                                                      B. sp. (p. 232)
2 Leaves once or twice pinnatisect, or if linear or narrow-spathulate, then entire or with a few irregular linear lobes or teeth   3

3	Leaves once or twice pinnatisect *B. nivalis* var. *nivalis* (p. 232)
3	Leaves linear or narrowly spathulate, entire or with a few irregular linear lobes or teeth *B. nivalis* var. *alpina* (p. 361)

4	Leaves with short glandular-septate indumentum, serrate distally *B. tenuiscapa* var. *tenuiscapa* (p. 361)
4	Leaves glabrous, entire 5

5	Achenes with minute glandular-septate hairs on the thick central region and the edge of the narrow thin margins, the pappus conspicuous; small stoloniferous plants to 7 cm high *B. stolonifera* (p. 362)
5	Achenes glabrous with short inconspicuous pappus; not as above 6

6	Base of the plant surrounded by the fibrous remains of old leaves; flowering stems usually with 0–2 bracts; phyllaries ± 12–18, broad and blunt or ± obtuse *B. scapigera* (p. 362)
6	Base of the plant not as above; flowering stems usually with 4–6 bracts; phyllaries ± 20–29, narrow, ± acute *B. obovata* (p. 363)

B. sp.

[295–6]   Tufted erect perennial herb ± 15–30(–50) cm high with a basal rosette of leaves and variable minute glandular indumentum; radical and lower cauline *leaves* broadly spathulate to obovate, ± 2–6(–8) × 1–3 cm, crenate or crenate-serrate, blunt, tapering to the base, the cauline leaves becoming cuneate and grading upwards into the small, toothed or entire linear ± acuminate bracts; *flowering stems* simple or sparingly branched above, glandular pubescent, densely so below the capitulum; *capitula* 2.5–4.5 cm in diameter including the pale violet or white ligules of the ray florets; *phyllaries* ± 6–9 mm long, narrowly elliptical to narrowly oblanceolate, acute or acuminate, densely glandular-pubescent on the outside; *achenes* ± 3–4 × 2.2–3 mm, obovate, flat, the body glabrous or with a few microscopic glandular hairs distally, the wings broad, subentire or irregularly and incompletely dissected, with minute antrorsely-appressed marginal cilia; *pappus* conspicuous; $2n = 18, 54^*$.

Distribution   See Notes below.

Notes and Habitat   Two forms of the *B. aculeata* species complex occur above the treeline: a small form up to about 15 cm high (diploid), which is mainly associated with feldmarks, and a more robust form (hexaploid), which is associated with heaths and tall alpine herbfields (Stace 1974). The complex is at present under revision.
References: Davis (1948); Smith-White *et al.* (1970)*; Stace (1974)*.

B. nivalis F. Muell., Trans. Phil. Soc. Vict. 1: 43 (1855)
var. *nivalis*   'Snow daisy'

[297–8]   Loosely tufted perennial herb ± 10–20(–30) cm high, glabrous or obscurely glandular-pubescent and often strongly tinged with purple towards the base;

61 *Lycopodium fastigiatum* (Mountain Clubmoss) has closely packed spore cases borne terminally in single or forked fertile spikes. It has a variable habit, from erect to trailing or decumbent in a wide range of habitats.

62 *Huperzia selago* (Fir Clubmoss) showing yellow spore cases in axils of leaf-like sporophylls, its upright habit and typical moist site near rocks.

61

62

63 *Grammitis armstrongii* is a dwarf fern usually found in shaded crevices of rock faces, and is uncommon in the alpine tract. Note the small entire leaves, in contrast to the divided leaves of most other species of alpine ferns. The spore cases (not visible) form a rounded mass on the underside of the leaves at maturity.

64 *Asplenium flabellifolium* (Necklace Fern), although widespread at lower elevations, is not common above the treeline. It prefers sheltered situations in rock crevices, etc. Note the fan-shaped pinnae and the curled tendril-like tip of the rachis at the top of the picture.

65 *Blechnum penna-marina* (Alpine Water-fern) is common and widespread in the alpine tract. Note that the pinnatisect leaves are of two types: sterile leaves in which the pinnae are close together, and more erect fertile (spore-bearing) leaves with narrower, more widely spaced pinnae. A group of fertile leaves can be seen on the upper left corner.

66 *Polystichum proliferum* (Mother Shield-fern) is the largest and, with *Blechnum penna-marina*, the commonest and most widespread fern at Kosciusko. It grows mainly among rocks.

63   64

65

66

67 *Cystopteris fragilis* (Brittle Bladder-fern) has a
world-wide distribution but is extremely rare at
Kosciusko above the treeline. It grows in damp
shady habitats.

68 *Podocarpus lawrencei* (Mountain Plum Pine), the
only native conifer above the treeline, is also the
longest-lived alpine species at Kosciusko. Its espalier
and rock-clinging habit enables the plant to benefit
from the locally warmer environments associated
with boulders. The roots have small nodules which
may fix atmospheric nitrogen.

69 Shoots of *Podocarpus lawrencei* showing the male
cones. Some of the cones have dehisced and the
yellowish pollen is visible on the dark green leaves.

70 Female cone of *Podocarpus lawrencei* showing the
swollen receptacle which has developed into a succulent berry-like structure above which is the hard
green seed.

67   68

69

70

71 *Hierochloe submutica* (Alpine Holy Grass) in flower, showing the sparse, drooping spikelets. Note also the rather broad leaves which are strongly scented when crushed.

72 *Agrostis meionectes* was described as recently as 1966. It is a delicate little grass, usually found in and around depressions in sod tussock grassland.

71
72

73

74

75

73 *Agrostis muellerana* at anthesis showing the typical dark contracted panicles; it is widespread above the treeline.

74 *Agrostis muellerana* (feldmark form). This form is a smaller, more compact plant. Dwarf ecotypes such as this are adapted to the extreme conditions of exposure characteristic of the feldmark.

75 *Agrostis parviflora.* This species prefers wet sites but occurs in a wide range of habitats.

76 *Deyeuxia affinis*. This rare little species, only recently described, is usually found near rocks in short alpine herbfield.

77 *Deyeuxia crassiuscula* is a rather coarse tufted grass with contracted spike-like panicles and broad, stiff, dark green, conspicuously ribbed leaves, the sheaths of which are often strongly tinged with reddish purple.

78 *Deyeuxia carinata* is superficially similar to *Deyeuxia crassiuscula* but is usually more slender and has narrower leaves.

79–80. *Trisetum spicatum* (Bristle-grass) typically shows three stages in the development of the panicle. The young panicle is contracted, then opens when the anthers develop and closes again at maturity. 79 shows the grass at anthesis, and 80 the panicle contracted at maturity. The stems of this widespread grass are usually velvety-pubescent.

81 The dwarf feldmark form of *Trisetum spicatum*.

76　79

77
78

80
81

82 *Deschampsia caespitosa* (Tufted Hair-grass) is of variable size but can be one of the largest grass species at Kosciusko. Typically it grows as isolated tussocks in wet sites and in shallow pools.

83–4 *Erythranthera australis* is the smallest alpine grass, sparsely distributed in colonies in short alpine herbfield and on other wet sites. Note the small heads and the stiff semi-erect leaves which sometimes show red banding on the sheaths. Until recently, this grass was thought to be restricted to the New Zealand Alps.

85 *Erythranthera pumila*. This dwarf alpine grass, first collected at Kosciusko in 1965, is possibly the rarest and most restricted grass in Australia. Like *E. australis*, it was previously thought to be endemic to the New Zealand Alps and is only known in Australia from the feldmarks on Mts Northcote and Lee.

82   83

84

85

86 *Chionochloa frigida* (Ribbony Grass) is the largest alpine grass and endemic to Kosciusko. It is particularly common on steep rocky slopes and along watercourses, where it forms dense communities. Like some other alpine species, *Chionochloa frigida* has become more common since the cessation of livestock grazing. It is closely related to other members of the genus in New Zealand. Lake Albina is in the background.

87–8 *Chionochloa* recolonizes steep rocky areas amongst heath and near watercourses.

86

87

88

89 *Danthonia alpicola* (Crag Wallaby-grass) in its typical habitat of steep, often inaccessible rock faces.
90 *Danthonia nudiflora* (Alpine Wallaby-grass) is particularly common in sod tussock grassland. Note the stiff sharp shiny leaves and the straight stems.
91 *Danthonia nivicola* (Snow Wallaby-grass) is the smallest member of the genus at Kosciusko. Note the fine leaves and the reddish stems. The latter often give the sod tussock grasslands in which this species is most common a reddish appearance when viewed against the light [48].
92 *Agropyron velutinum* (Velvet Wheat-grass) can be distinguished by its wheat-like heads and the pubescence on the upper part of the stems.
93 *Poa saxicola* (Rock Poa), a widespread snow grass, differs from the other species of *Poa* in its plump spikelets and contracted panicle which often droops at maturity. Although often found in rocky sites, this species can also form colonies in open herbfield.

89   90

91

93

92

94

95
97

94 *Poa costiniana* (Prickly Snow Grass) is a wide-spread alpine snow grass especially common in damp sites in which it may form dense pure swards. It tends to be the most robust species of *Poa* above the treeline. Note the coarse stiff shiny green leaves.

95 *Poa fawcettiae* (Smooth-blue Snow Grass) is a characteristic snow grass of the tall alpine herbfields but it also occurs in other communities. As noted in the botanical descriptions, *Poa fawcettiae* and *Poa costiniana* are not always easy to distinguish above the treeline and their field distribution is not yet fully understood.

96 *Poa hiemata* (Soft Snow Grass), also common in tall alpine herbfields, is more delicate than the other alpine species of *Poa*, with finer and softer leaves and relatively small spikelets.

97 *Scirpus crassiusculus* showing the relatively large solitary and terminal spikes. Like most other members of the family Cyperaceae at Kosciusko, it prefers wet sites.

98 *Scirpus aucklandicus* showing the small solitary spikelet enfolded at the base by the erect floral bract.

99 *Scirpus subtilissimus*, a dwarf rhizomatous species, has 1–3 small spikelets along the stems and forms short dense swards.

96

98

99

100 *Scirpus habrus*. Dwarf plants of this species sometimes closely resemble *S. subtilissimus* but it may be distinguished by its tufted habit and larger size. A fruiting head of *Oreomyrrhis ciliata* is on the left.

101 *Scirpus montivagus* is a dwarf mat-forming species, sometimes forming extensive swards. The spikelets are larger than those of *S. subtilissimus* and are almost sessile among the leaf bases.

100

101

102

102 *Carpha nivicola* (Broad-leaf Flower-rush) showing the large feathery clusters of spikelets subtended by leaf-like bracts. It is fairly widespread in bogs, fens and other wet sites.
103 *Carpha alpina* (Small Flower-rush) is more delicate in appearance than *Carpha nivicola*, found in similar sites. It has finer leaves and smaller clusters of spikelets often in more than one group along each stem.

103

104 *Schoenus calyptratus* (Alpine Bog-rush) showing the reddish purple leaf sheaths and partly concealed spikelets. Note the long filamentous stigmas. This dwarf species often forms a turf in short alpine herbfield.

105 *Oreobolus distichus* (Fan Tuft-rush). The leaves of this species tend to be arranged in two rows. Most of the glumes have fallen off to reveal the solitary nuts subtended by the tooth-like hypogynous scales. 106–8 *Oreobolus pumilio* (Alpine Tuft-rush) has short flattened leaves which are not arranged in rows and is also distinguished from *Oreobolus distichus* by the shorter nuts which do not extend above the scales. In 107 most of the glumes are still in position whereas 108 shows the mature spikes from which the glumes have fallen to reveal the brown nuts enclosed by small tooth-like scales. *Oreobolus pumilio* grows both as small cushions and as extensive swards. Sometimes the cushions die off in the middle and the margins continue to grow outwards to form ring-like patterns [106].

104

105

106

107

108

109 *Uncinia sinclairii.* This rhizomatous species has only recently been collected at Kosciusko and could possibly be an introduction from New Zealand where it was previously thought to be endemic. The leaves are broad, relatively short and erect and are over-topped by the pale brown flowering heads. Note the hooked seeds which assist in dispersal.

110 *Uncinia* sp. This undescribed species is the most common *Uncinia* above the treeline and is found in tall alpine herbfield and sod tussock grassland.

111 *Carex cephalotes* is easily distinguished from the other species in the area by the compact solitary spike. At maturity the female part of the spike opens and the utricles become spreading or reflexed.

112 *Carex breviculmis* is a fairly common sedge of dry habitats. Like *Carex jackiana* its spikes are almost hidden among the base of the leaves; however, the utricles of *C. breviculmis* are covered with minute hairs.

113 *Carex jackiana* also carries the spikes hidden among the base of the leaves but the utricles are glabrous and it is found in wet areas usually around the margins of bogs.

114–15 *Carex gaudichaudiana* is the most common and widespread of the alpine sedges. The terminal spike consists of male flowers, whereas the lower spikes consist mainly of female flowers. 115 shows young plants with the whitish stigmas extruded.

109

110

111

112

113

115

114

116

117

118

120

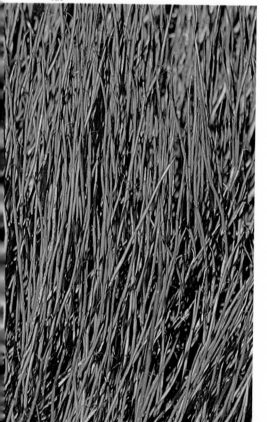

119

121

116 *Carex gaudichaudiana* fen.
117 *Carex hypandra* is generally similar in appearance and habitat to *Carex gaudichaudiana* and is often found associated with it. It is distinguished by its usually darker, shorter and more congested spikes, the terminal one of which is female in the upper part.
118 *Carex echinata* (Star Sedge) has small star-shaped spikes which give the plant its common name. It is fairly uncommon above the treeline in moist sites.
119 *Carex curta*, another common sedge of wet habitats has compact yellowish green spikes.
120 *Carex hebes* is the commonest sedge of dry habitats. Note the shining green leaves and reddish brown spikes carried on long thin stems.
121 *Empodisma minus* (Spreading Rope-rush) is common in wet sites, especially in *Sphagnum* bogs. It has thin wiry stems with rudimentary leaves and inconspicuous flowers which are hidden in the uppermost sheaths.

122–3 *Juncus antarcticus* (Cushion Rush), a dwarf alpine rush, is characterized by its short very stiff channelled leaves and terminal few-flowered heads. It forms dense turf or cushions in wet sites.

124 *Juncus falcatus* (Sickle-leaf Rush) is the largest native *Juncus* above the treeline and can be recognized at maturity by the large dark globular flower clusters.

125 *Juncus falcatus* shown at anthesis when the conspicuous pale mauve stigmas and yellowish anthers are particularly attractive.

126 *Juncus* sp., a small and inconspicuous species recently recorded from Kosciusko, is still undescribed. It has small green clusters of spikelets and soft leaves with faint transverse partitions only visible when the leaves are flattened.

122
123

124

125

126

129 *Luzula novae-cambriae*, a robust species of heath and other rocky sites, can be recognized by its open clusters of flowers and the broad soft leaves. The long white hairs on the leaf margins can also be seen on other members of the genus. This species is believed to form hybrids with *L. australasica*, a species found in bogs, where their habitats adjoin. The photograph shows the plant at an early stage of flowering.

127–8 *Luzula acutifolia* subsp. *nana* is a distinctive dwarf mat-forming plant endemic to Kosciusko with short stiff stems and leaves and blackish heads. Its main habitat is short alpine herbfield. The photographs show the plant in fruit and in flower.

127

128

129

130 *Luzula atrata* is a slender rhizomatous species with blackish heads and often reddish stems; it usually occurs in large colonies and prefers wet habitats.

130

131 *Luzula australasica* is a species of *Sphagnum* bogs.
Note the flat leaves, the broad leaf-like bracts and
the elongated clusters of flowers.
132 *Luzula oldfieldii* subsp. *dura* is usually a robust
plant with stiff leaves and large globular heads found
in dry feldmark and other rocky habitats. It is also a
vigorous colonizer of eroded alpine herbfield areas.
133 *Luzula alpestris* is a densely tufted species with
stiff leaves and stems, differing from *Luzula atrata* in
its brownish heads and in its preference for sod
tussock grassland.

131
132
133

134–6 *Dianella tasmanica* (Tasman Flax-lily) has large broad leaves, blue flowers and attractive green fruits which turn bluish violet when ripe. It is uncommon above the treeline.

137 *Herpolirion novae-zelandiae* (Sky Lily) usually forms patches or grass-like swards; the pale lilac-blue to whitish flowers are large for such a small plant but short-lived. It is rare above the treeline.

134
135
136

138–9 *Astelia alpina* has stiff short-pointed leaves which give this species its common name of Pineapple Grass. It grows in wet sites, typically *Sphagnum* bogs. 138 shows the exserted male inflorescence.

137

139

138

LILIACEAE

140–3 *Astelia psychrocharis* (Kosciusko Pineapple Grass) has broader leaves which are more attenuate at the apex. Both this and the previous species are dioecious, i.e. the male and female flowers are borne on separate plants. 141 shows the contracted male inflorescences, 143 the 3–celled orange-yellow fruit. 142 *Astelia psychrocharis* growing with *Ranunculus gunnianus*.

140

141

142

143

144

145

146

147

144 *Caladenia lyallii* (Mountain Caladenia),
although fairly common around *Sphagnum* bogs in
the subalpine tract, is very rare above the treeline.

145–6 *Prasophyllum alpinum* (Alpine Leek-orchid)
has small green to reddish brown flowers and is
common above the treeline. Insects are attracted to
the nectar of the fragrant flowers.
147 *Prasophyllum suttonii* (Mauve Leek-orchid) can
be distinguished by its larger white and mauve-
tinged flowers. This orchid is also common above
the treeline.

148 *Orites lancifolia* (Alpine Orites), a common alpine shrub, has showy creamy white spikes of flowers, and the tough boat-shaped fruits typical of many species of the family Proteaceae can also be seen.

149 *Grevillea australis* (Alpine Grevillea) is a widespread alpine shrub with characteristic cream-coloured 'spidery' flowers. The leaves are sharp-pointed but otherwise vary greatly in shape from almost needle-like to broadly elliptical or obovate.

150 This form of *Grevillea australis* has broader leaves. Note the reddish clusters of buds and the ovoid fruits. Two fruits have matured to a purplish colour and split open and a third less mature fruit is still green.

151–2 *Grevillea victoriae* (Royal Grevillea) is a larger shrub with spectacular clusters of red flowers and large leaves. The rust-coloured buds are shown in 151. This variable species is rare above the treeline, mainly on the steep, west-facing slopes of the Main Range.

148   149

150

151

152

153 *Exocarpos nanus* (Alpine Ballart) is a dwarf shrub in which the flowers are inconspicuous and the leaves are reduced to small scales on the flattened greenish branchlets. The flower stalk matures into a bright red succulent structure which supports the smaller nut-like fruit, and superficially resembles the nut-like seed of *Podocarpus lawrencei* on its swollen red receptacle. *Exocarpos nanus* is a root parasite.

153

154

155

154–7 *Neopaxia australasica* (White Purslane) is one of the main species of snow patch sites, and also occurs in other semi-bare, usually damp places. It has a stabilizing and soil-building function similar to that of *Caltha introloba*, *Plantago glacialis* and other associated snow patch plants. The flowers are usually white and the leaves bright green, but pink-flowering forms [156] and bluish-leaved forms [157] are not uncommon.

156

157

158 *Scleranthus biflorus* (Two-flowered Knawel) forms compact green cushions of small, crowded pointed leaves. The small hard flowers occur in terminal pairs on the stems.
159–60 *Scleranthus singuliflorus* (One-flowered Knawel) is usually a more lax species than *Scleranthus biflorus* and there is only one flower on each stem.

< 158   159
          160

CARYOPHYLLACEAE

161 *Colobanthus affinis* (Alpine Colobanth) is a small tufted herb with soft grass-like leaves; the flowers and fruiting capsules (shown open) are carried on long stalks.

162 *Colobanthus pulvinatus* (Hard Cushion-plant), one of the few true cushion-plant species at Kosciusko, has stiff, very prickly leaves among which are numerous flowers. It occurs mainly on wind-swept feldmark sites.

163–4 *Colobanthus nivicola* (Soft Cushion-plant), closely related to *C. pulvinatus*, is distinguished by its softer, slightly narrower leaves and preference for moister snowier sites. 163 shows the moss-like cushions of this species coalescing to form a dense cover on bare eroded areas.

161
162

163
164

Caltha introloba (Alpine Marsh-marigold) is one of the most specialized alpine species and grows in wet cold snow patch habitats. It develops under the winter snow cover and plays an important role in building and stabilizing soil. The sweetly perfumed flowers, although distinctive, show considerable variability in size, shape and colour.

166 *Caltha* showing different stages of development. The flowers start to open under the snow and the new leaves develop among the decayed remains of the last season's growth.

167 *Caltha* is often inundated by snow-melt waters.

165 *Stellaria multiflora* (Rayless Starwort) is very rare and quite dwarfed above the treeline, and grows on bare rocky sites.

165

166
167

168–71 Variation in the shape and colour of *Caltha* flowers. Note also the bright green leaves, with the characteristic twin lobes at the base.
172 *Caltha* in fruit.

168   169

170

173–5 *Ranunculus anemoneus* (Anemone Buttercup), with its large white anemone-like flowers and
dissected stem-clasping upper leaves, is the largest
buttercup at Kosciusko and one of the most beautiful
in the world. It is related to some of the spectacular
New Zealand alpine species.

173

174

175

176 *Ranunculus anemoneus* in flower, with *Aciphylla glacialis* (stiff pointed leaves). Since the Kosciusko summit area was protected from grazing in 1944 these two species together with other palatable plants have made a strong recovery.

176

177

177–8 *Ranunculus millanii* (Dwarf Buttercup) has white to cream-coloured flowers and is the smallest alpine buttercup, forming short mats in wet sites. It can grow for considerable periods submerged in water [178].

179–80 *Ranunculus muelleri* (Felted Buttercup) has elliptical pointed leaves toothed towards the apex and covered with long hairs. Note the beaked fruits which are characteristic of this and other species of *Ranunculus*.

178

179

180

181

182

183
184

181–2 The dwarf *Ranunculus muelleri* var. *brevicaulis*
is restricted to *Epacris–Chionohebe* feldmark whereas
the variety *muelleri* has a wider distribution. There
are many intermediate forms between the two
varieties where their habitats overlap.

183 The windswept feldmark, shown here along the
Main Range, characterized by the dwarf and prostrate clumps of *Epacris microphylla* (Coral Heath) is
also the main habitat of many other dwarf and hardy
plants including *Ranunculus muelleri* var. *brevicaulis*
[181–2], *Chionohebe densifolia* [272–3] and *Euphrasia
collina* subsp. *lapidosa* [270–1].

184 *Ranunculus dissectifolius* has divided leaves with
linear-lanceolate segments that are covered with
spreading hairs, and the flowers are carried on hairy
stems.

185

187

185 *Ranunculus graniticola* (Granite Buttercup) has broadly dissected leaves and bright yellow flowers which can have either greenish or dark-coloured centres.

186 *Ranunculus gunnianus* (Gunn's Alpine Buttercup) has bright golden yellow flowers and very finely dissected bright green leaves, the individual segments of which are subterete.

187 *Ranunculus niphophilus* (Snow Buttercup) is endemic to the Kosciusko area and is distinguished by its crowded, shiny, dark green leaves. It forms extensive colonies below snow patches and in seepage areas where it flowers profusely.

188–9 *Tasmannia xerophila* (Alpine Pepper) has bright green leathery strap-shaped leaves, terminal clusters of creamy flowers on long stalks, and purplish black fruits at maturity [189]. The leaves have a sharp, peppery taste when chewed. Male and female flowers occur on separate plants; the male flowers are shown in 188.

186

188

189

190 *Cardamine* sp. This is the common form at Kosciusko which occurs in herbfields and heaths.
191 *Cardamine* sp. This robust form with larger flowers and leaves and spreading rootstock is more restricted in distribution and forms extensive colonies in rocky sites near late snow areas, usually on the margins of streams and lakes.

192 *Drosera arcturi* (Alpine Sundew) has conspicuous solitary white flowers and strap-shaped leaves with large sticky glandular hairs which trap insects. It is the only insectivorous species above the treeline.
193 *Crassula sieberana* is an inconspicuous dwarf annual herb of bare rocky sites, with reddish flowers borne in the axils of the small fleshy leaves.
194 *Acaena* sp. is easy to recognize by its pinnate leaves and globular flower heads. The socks of summer visitors, covered with the spiny seed cases, testify to the effective seed dispersal mechanism of this plant.

190

191

193

192 194

195 *Alchemilla xanthochlora* (Lady's Mantle) has
large fan-like leaves and clusters of greenish yellow
flowers. The shining, dark green pinnate leaves and
immature green globular flower heads in the background are of *Acaena* sp.
196 *Hovea purpurea* var. *montana* (Alpine Hovea),
an early-flowering leguminous shrub, has purplish
or occasionally white pea-shaped flowers and oblong
discolorous leaves. It is not common above the
treeline but is very common in the subalpine tract.
197-8 *Oxylobium ellipticum* (Golden Shaggy-pea), a
widespread shrub of heaths, has relatively short
pointed leaves and bright orange-yellow flowers.
The hairy pods are shown in 197.
199 *Oxylobium alpestre* (Mountain Shaggy-pea),
common at lower altitudes, appears to be mainly
restricted to the western slopes of the alpine tract.
The leaves have stipules and are generally larger
than those of *Oxylobium ellipticum*.

195   196

197
198 199

200 *Geranium antrorsum* (Rosetted Crane's-bill) has semi-orbicular to wedge-shaped leaves. This photograph shows the characteristic *Geranium* fruits which have split open to shed the seeds.
201 *Geranium potentilloides* var. *potentilloides* showing the deeply dissected orbicular leaves and solitary white flowers on long stalks.

202–3 *Phebalium ovatifolium* (Ovate Phebalium), a common aromatic shrub restricted to the Kosciusko area, has a profusion of white flowers [202]. It has a characteristic flat-topped crown when growing in the open. In addition to the oil glands which cover this plant, the branchlets and undersides of the leaves are thick with dense scurfy scales; 203 shows the fruit.
204 *Stackhousia pulvinaris* (Alpine Stackhousia) forms compact mats of fleshy, shiny green leaves dotted with sweetly perfumed, creamy yellow star-shaped flowers.

200
201

202
204
203

205 *Viola betonicifolia* (Showy Violet) is the
common native violet above the treeline. Flower
colour varies from almost white to pale purple or
deep violet.

206 *Hymenanthera dentata* var. *angustifolia* is a
much-branched, prostrate spiny shrub of dry rocky
habitats. A branch of this plant has been laid on its
side to show the small yellowish flowers which are
normally hidden along the undersides of the branchlets. The subglobular berries are an attractive dark
purple when ripe.

205

206

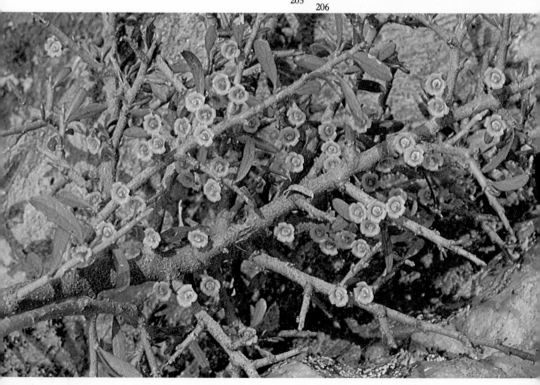

207–8 *Drapetes tasmanicus* is a small creeping plant of
exposed feldmark sites, characterized by stiff
crowded hairy leaves and white flowers [207]; the
small brown fruits are shown in 208.

207

208

209 *Pimelea axiflora* var. *alpina* is a small erect or
ascending subshrub with the creamy white flowers
in axillary clusters along the stems.
210 The alpine form of *Pimelea ligustrina* is the
largest and most spectacular *Pimelea* at Kosciusko
and ascends just above the treeline. Terminal flower
clusters are surrounded by large bracts usually strongly tinged with magenta when the plant is in fruit.

209

210

211–12 *Pimelea alpina* (Alpine Rice-flower) is a prostrate subshrub with small leaves and masses of sweetly perfumed flowers varying in colour from white to deep pink.

211

212

213
214

216

213 This view of Mt Clarke from near Charlotte's Pass shows many yellowish patches of flowering *Kunzea muelleri* heath amongst and beyond the Snow Gums.
214 *Kunzea muelleri* (Yellow Kunzea) forms extensive low heaths and has conspicuous flowers [37].

215 *Baeckea gunniana* (Alpine Baeckea) is common in wet boggy sites. Its flowers have fewer stamens, usually 5(–7), and the leaves are smaller and narrower than in *Baeckea utilis*.
216–17 *Baeckea utilis* (Mountain Baeckea) prefers damp rocky sites and watercourses. The flowers have more stamens than *Baeckea gunniana*, usually 8 (7–10), and the leaves are larger and flatter.

215

217

ONAGRACEAE

218–19 *Epilobium tasmanicum* (Snow Willow-herb) is a prostrate creeping species of wet gravelly sites. 219 shows the small white flowers with notched petals and reddish capsules. The opened capsules on elongated stalks are shown in 218.

218
219

220

220–1 *Epilobium gunnianum* (Gunn's Willow-herb) is an erect species usually found in dense colonies. Like other members of this genus at Kosciusko, it has opposite leaves and long capsules which at maturity split open from the top to release the seeds.

222–3 *Epilobium sarmentaceum* (Mountain Willow-herb) is a tall species like *Epilobium gunnianum* but is much less common and occurs in rock outcrops in drier sites. The stalks of the capsules are longer than those of *Epilobium gunnianum* and the flowers tend to be smaller.

221
222

223

HALORAGACEAE/UMBELLIFERAE

224–5 *Myriophyllum pedunculatum* (Mat Water-milfoil) is a succulent creeping species which forms mats on damp sites. It is dioecious, with the female [224] and male [225] flowers on separate plants.
226 *Gonocarpus micranthus* subsp. *micranthus* (Creeping Raspwort) is a dwarf creeping herb of damp sites with small orbicular, almost sessile leaves and red clusters of flowers and fruits.
227 *Gonocarpus montanus* is a larger plant with longer, pointed leaves.
228–9 *Oreomyrrhis pulvinifica* (Cushion Carraway) is the smallest species of this genus at Kosciusko, typically found below snow patches and in similar wet sites. It has small white flowers and reddish fruits.

224
225

226

227

228
229

230 *Oreomyrrhis ciliata* (Bog Carraway), seen here in fruit, has finely dissected almost glabrous leaves which have a strong sickly sweet odour, and slender bristly flower stalks.

231–2 In *Oreomyrrhis brevipes* (Rock Carraway) the stems are short and branched and the leaves velvety. Many intermediate forms occur between this and *Oreomyrrhis eriopoda*. 232 shows a small fruiting plant growing in the feldmark.

<230   231   232

233 *Oreomyrrhis eriopoda* (Australian Carraway) is
the commonest and most variable species of this
genus at Kosciusko, distinguishable by its long flower
stalks which arise from the basal rosette of hairy
leaves. Many intermediates are found between this
species and *Oreomyrrhis brevipes*.
234 *Aciphylla simplicifolia* (Mountain Aciphyll) is a
small species with slender segmented leaves. The
specimen shown here is a male plant.
235–7 *Aciphylla glacialis*, appropriately named
Mountain Celery because of the thick fleshy stems, is
a dioecious species. 235 and 237 show the delicate
male flowers and 236 shows a female plant in fruit.

233   234

235

237

236

238 *Aciphylla glacialis* growing in a mat of *Celmisia* foliage.

239–40 *Gingidia algens*, a low-alpine and subalpine species, has parsnip-like leaves and small white flowers carried in open clusters [239]. The flower stalks elongate at maturity. It is usually found in *Sphagnum* bogs, and in 240 the fruiting plant is surrounded by the wiry stems of *Empodisma minus*. *Gingidia* has a strong aniseed odour when crushed.

238   239   240 >

241 *Oschatzia cuneifolia* (Wedge Oschatzia). This rare species has long narrowly wedge-shaped leaves toothed at the top, and small white flowers carried on long slender stalks.

242 *Diplaspis hydrocotyle* (Stiff Diplaspis) has dark green, rounded shiny leaves, inconspicuous flowers and terminal clusters of fruits borne on swollen light green stalks.

243 *Schizeilema fragoseum* (Alpine Pennywort), seen here in fruit, has small pale green clusters of flowers hidden among the rounded lobed leaves. It is usually found growing in dense patches in moist shady sites around the bases of rocks.

244 *Dichosciadium ranunculaceum* var. *ranunculaceum* is a rosette plant with a thick taproot, often forming extensive patches, with large palmately lobed hairy leaves and numerous white flowers.

243

241

244

245

245–6 *Richea continentis* (Candle Heath) has large prickly leaves and a terminal spike-like inflorescence of showy creamy flowers borne on reddish stems. It is usually associated with *Sphagnum* moss and peat. 245 shows the swollen corolla, which does not open as in other flowers, and the broad pointed bracts, the uppermost of which have hooked tips.

245

246

The *Epacris* species, like the ecologically related
European heaths, have small leaves and generally
grow on wet or poor soils. The four species of *Epacris*
at Kosciusko, which are all sweetly perfumed, are
sometimes similar in appearance but can be distinguished in the field by their leaf shapes and often
by differences in habitat.

247 A raised bog community with *Richea continentis*
and *Epacris paludosa* near the Ramshead Range.
Damage by past fires and trampling by livestock has
caused many alpine bogs to erode down to the
underlying rock or gravel; some such areas are now
in various stages of recovery, with short alpine herbfield species colonizing the wet stony surface and
banks of *Sphagnum* moss slowly extending from the
edges.

248 *Epacris paludosa* (Swamp Heath) is the tallest
species of *Epacris* above the treeline, with the largest
leaves and flowers and a preference for watercourses
and boggy or wet bouldery habitats. Note the large
prickly lanceolate leaves.

249–51 *Epacris glacialis* has leaves of rhomboid shape
and conspicuous reddish brown bracts and sepals
[249]. 250 and 251 show the plant in two characteristic habitats, the first in *Sphagnum* bog growing with
the wiry rush *Empodisma minus*, and (overleaf)
growing close over rocks.

<247 248
249

250

251 *Epacris glacialis* growing close over rocks.

252 *Epacris microphylla* (Coral Heath) grows in a
variety of habitats including exposed feldmark. It has
small heart-shaped leaves and clusters of white
flowers with inconspicuous bracts and sepals.

253 In feldmark habitats the *Epacris* shrubs are
characterized by simultaneous death and regeneration on the exposed and leeward sides respectively.

254 *Epacris petrophila* (Snow Heath) showing the
narrow sessile overlapping leaves. This species is also
found in a few feldmarks as well as in heaths and bogs.

251 252

253 254

255

256

255–6 *Pentachondra pumila* (Carpet Heath), a mat-forming subshrub, is one of the most attractive alpine species because of the profusion of succulent crimson fruits and white star-like flowers. Note that some of the old flowers are retained on the fruits.

257–8 *Leucopogon montanus* (Snow Beard-heath) is a low shrub which often carries the current season's flowers with the previous season's fruit. The leaves have a characteristic parallel venation on the pale underside. 257 shows the male flowers and 258 the female flowers and fruits. Unlike most Beard-heaths, the corolla lobes of this species are not hairy.

257
258

259

261

262

262–3 *Prostanthera cuneata* (Alpine Mint-bush) is a strongly aromatic shrub with masses of white or pale violet flowers which form a beautiful contrast on the background of dark green densely gland-dotted leaves.

259–60 *Gentianella diemensis* (Mountain Gentian) has beautiful white crocus-like flowers with purplish veins. This species forms extensive communities in sod tussock grassland and usually flowers rather late in the season.

261 *Myosotis australis*. The rare Kosciusko form of this species has larger flowers than forms from lower altitudes.

260

263

LABIATAE/SCROPHULARIACEAE

264 Like many of the shrubs in the Kosciusko heaths, *Prostanthera cuneata* is often found creeping over rocks.

265 *Euphrasia alsa* (Dwarf Eye-bright), the smallest species of this genus at Kosciusko and one of the relatively few annuals above the treeline, has small rather fleshy leaves and short racemes of white or pale mauve flowers with dark veins.

266–7 *Euphrasia collina* subsp. *glacialis* is a white-flowered plant with a preference for wet sites. Mass flowering of *Euphrasia collina* subsp. *glacialis* early in the season in a typical moist habitat near snow patches.

264
265

268–9 *Euphrasia collina* subsp. *diversicolor* is the largest and most widespread *Euphrasia* in the alpine tract. The flowers are very showy and vary in colour from shades of violet or lilac to white. It is found in drier sites than the subspecies *glacialis*.

270–1 *Euphrasia collina* subsp. *lapidosa* is a low–growing feldmark subspecies, with the leaves crowded together on the stems. The colour of the flowers sometimes resembles that of *Chionohebe* with which it grows.

269

270

271

272–3 *Chionohebe densifolia* is the only Australian mainland representative of a genus which is more common in New Zealand. It forms dense mats in feldmark with thick leaves overlapping in rows and showy mauve to white flowers.

274 *Veronica serpyllifolia* (Thyme Speedwell) is an inconspicuous herb, rare above the treeline, with opposite leaves and small dark-veined pale mauve to violet flowers.

272
273

274>

275 *Plantago glacialis* (Small Star Plantain), with patches of *Ranunculus niphophilus* in flower, forms a characteristic mat of short alpine herbfield below this snow patch. The *Plantago* plants accumulate as peats, some of which are about 2500 years old. The age of these peats indicates the approximate duration of the present-day alpine climate with its associated snow patch regime.

276 *Plantago glacialis* (Small Star Plantain) is a species of wet snow patches and similar sites, and forms extensive mats in which the star-shaped rosettes tend to be regularly spaced. The reddish brown flowers and fruits, only a few to each cluster, are carried on short stalks. This is the smallest species of the genus at Kosciusko and varies in size and in the shape and degree of hairiness of the fleshy leaves.

276

277

278

277–8 *Plantago muelleri* (Star Plantain) is more robust than *Plantago glacialis*, with very thick fleshy shining leaves. 277 shows this species in flower, with the heads almost hidden among the leaf bases, and 278 shows a fruiting plant with elongated scapes.

279 *Plantago alpestris*, found in damp sites, has slightly hairy leaves with the midvein of the leaf more conspicuous than the lateral veins.

280 *Plantago euryphylla* is the largest of the alpine plantains, with broad pubescent strongly ribbed leaves and long flower heads. It is not common above the treeline and is found in drier sites than are the other species of *Plantago*.

281 *Asperula gunnii* (Mountain Woodruff) is fairly common as a minor component of taller communities. The photograph shows a female plant and the two-branched styles with conspicuous stigmas.

282 *Asperula pusilla* (Alpine Woodruff) is much less common than *Asperula gunnii* above the treeline and has smaller, narrower, hispid-pubescent leaves.

283–5 *Coprosma* sp., a spreading mat-forming shrub of snowy feldmark sites, is characterized by its bright orange-red fruits and dioecious flowers. Female flowers and last season's fruits [284] and male flowers with long exserted anthers [285] are shown.

279
280

281
282

283

284

285

286 *Nertera depressa* is a dwarf creeping herb of damp sites. It has tiny inconspicuous flowers and succulent red fruits, and is uncommon above the treeline. The small thick ovate leaves are shown here almost hidden among mosses and other plants (e.g. *Ranunculus muelleri*).

287 *Wahlenbergia ceracea* (Waxy Bluebell), the common bluebell above the treeline, has pale blue to pale violet flowers and mostly alternate leaves.
288 *Wahlenbergia gloriosa* (Royal Bluebell) has deep purplish violet flowers and the leaves are more crowded and mostly opposite with more undulate margins than in *Wahlenbergia ceracea*. It is rare above the treeline but quite common in the subalpine tract.

286

287

289 *Pratia surrepens* (Mud Pratia) is a small creeping herb of wet sites with white or very pale violet flowers and semi-orbicular leaves. It is mostly found in small depressions in sod tussock grassland.

290 *Goodenia hederacea* var. *alpestris* is very marginal above the treeline although quite common in the subalpine tract. It has creeping stems and coarsely serrate or dentate leaves which are white and appressed-tomentose on the underside.

288
289

290

291–2 *Stylidium graminifolium* (Grass Trigger-plant) has tufts of stiff grass-like leaves and spectacular deep pink to purplish red flowers which are peculiarly adapted to pollination by insects. When the base of the irritable column or 'trigger' is touched by an insect in search of nectar it springs downwards and dusts the insect's back with pollen before it proceeds to the next flower. 292 shows the trigger and the linear appendages in the throat of the corolla tube.

293 Massed flowering of *Stylidium graminifolium* in a damp site below Etheridge Range.

294 *Lagenifera stipitata* is a widespread but inconspicuous minor herb.

295–6 *Brachycome* sp. is a widespread alpine daisy characterized by its pale violet to white flower heads. A small ecotype [295] occurs in feldmark.

291

292

293

294

296

295

299 *Brachycome nivalis* var. *alpina* can easily be distinguished from the variety *nivalis* by its narrow leaves which are entire or irregularly toothed. It is a stoloniferous plant found in wet areas.

300 *Brachycome tenuiscapa* var. *tenuiscapa* has a stoloniferous habit and basal rosettes of crenate-serrate leaves and is usually found growing in extensive patches. It is not common above the treeline.

297–8 *Brachycome nivalis* var. *nivalis* (Snow Daisy) has finely dissected leaves and prefers steep rocky habitats.

297
299

300

298>

301
302
303

301 *Brachycome stolonifera*, a dwarf glabrous stoloniferous daisy, is characteristic of short alpine herbfields, fens and other damp sites.

302 *Brachycome scapigera* (Tufted Daisy) superficially resembles *Brachycome obovata* but may be distinguished by the fibrous remains of old leaves around its base and its preference for drier sites.

303 *Brachycome obovata* has showy white or pale mauve flowers, and prefers wet sites.

304–6 *Erigeron setosus*, a dwarf species, has small, very hairy leaves, and white flowers growing close to the ground. The stalks elongate as fruits develop [304].

304
305

306

307 *Erigeron pappocromus*. This photograph shows the common form (form B) of this species above the treeline. It has viscid leaves sparsely covered with very short stiff hairs. There are two other less common forms above the treeline, one with pubescent leaves and a smaller form with long slender rhizomes.

308–11 *Celmisia* sp. (Silver Snow Daisy) occurs in a variety of forms and habitats throughout the Kosciusko alpine region. Its major occurrence is in the *Celmisia–Poa* tall alpine herbfield alliance where the silvery foliage and masses of white flowers form extensive and continuous carpets, a major and distinctive feature of the alpine landscape.

307   308

309

310

311

312 *Olearia algida* (Alpine Daisy-bush) is marginal above the treeline in heaths. The small appressed leaves of this shrub resemble those of *Helichrysum hookeri* but flower heads of the latter lack the conspicuous white petal-like ligules of the outer florets.

313 *Olearia phlogopappa* var. *subrepanda* is a small floriferous shrub with small greyish hairy leaves, which is fairly common in rocky sites.

314 *Olearia phlogopappa* var. *flavescens* is rather similar to the variety *subrepanda* but differs in the larger leaves and the longer flower stalks. It is one of the most spectacular alpine daisy bushes.

315 *Gnaphalium umbricola* (Cliff Cudweed), a species of damp rock crevices, is extremely rare above the treeline and needs rigorous protection.

312
314

313

315>

316 *Gnaphalium argentifolium* (Silver Cudweed) is the commonest cudweed above the treeline, usually forming small patches or mats.

317 *Gnaphalium nitidulum* (Shining Cudweed) is a small silvery-leafed mat–forming species superficially similar to *Gnaphalium argentifolium* but differing in the larger solitary flower heads on very short stalks and the coarser hairs on the leaves.

318 *Gnaphalium fordianum* is closely related to *Gnaphalium argentifolium* [316] but is a larger plant with longer, more numerous flower heads on each stem. The numerous flower heads of this species are at first congested in a terminal cluster, and their stalks usually elongate at maturity to give a more open inflorescence.

316 317

319–20 *Ewartia nubigena* (Silver Ewartia) usually forms extensive mats or low hummocks and the silvery grey tomentum of the leaves often becomes brownish with age.

318

320

319

COMPOSITAE

321–2 *Parantennaria uniceps*, one of the smallest alpine daisies, has stiff crowded sharp-pointed leaves and forms patches or dense swards mainly in short alpine herbfield. 321 shows a male plant with broad sessile flower heads. The female plant shown in 322 has narrower flower heads which are raised on short stalks when in fruit.

321

322

323

323–5 *Helipterum albicans* subsp. *alpinum* (Alpine Sunray), a spectacular everlasting of exposed, alpine herbfield [323] and feldmark [325] sites, has large white papery flower heads with orange-yellow centres and silvery woolly leaves. 324 shows extensive patches in and adjacent to feldmark areas along the Main Range.

324

325

326 *Helipterum anthemoides* (Chamomile Sunray), a white-flowered everlasting, is marginal in the alpine tract as small colonies in tall alpine herbfields.

327 *Helichrysum scorpioides* (Button Everlasting) with its golden yellow flower heads is locally common above the treeline.

328 *Helichrysum hookeri* (Scaly Everlasting) is an aromatic shrub with congested clusters of sessile flower heads and very small leaves closely appressed to the stems. It is marginal above the treeline.

329 *Helichrysum secundiflorum* (Cascade Everlasting) is a very aromatic shrub with clusters of creamy white flower heads borne along the main stems. The leaves are somewhat sticky with webby hairs on the upper surface and dense grey tomentum on the underside.

330 *Helichrysum alpinum* (Alpine Everlasting) is a small shrub with terminal clusters of flower heads with distinctive reddish phyllaries; the branchlets and underside of the young leaves are covered with golden yellow tomentum.

326
327

329

330

328

331–2 *Leptorhynchos squamatus* (Scaly Buttons) is a variable species, two forms of which are shown in 331. It prefers semi-bare habitats and is an effective colonizer.

333–4 *Podolepis robusta* (Alpine Podolepis) has large, rather fleshy leaves and conspicuous clusters of golden yellow flower heads.

335 *Craspedia leucantha* is distinguished by its whitish flower heads and almost glabrous dark green leaves. It is found in wet areas below snow patches and on wet banks of streams.

333
334

335

336 *Craspedia* sp. A is the smallest species above the treeline with narrow silvery leaves and creamy white flower heads. It is usually found on wet mud and in small pools.

336

338

337

337–8 *Craspedia* sp. B has discolorous leaves; the upper surface is greenish and densely covered with small sticky glandular hairs, the lower surface whitish with appressed woolly hairs. It is mostly found in rocky sites.

339 *Craspedia* sp. C is a robust species with silvery or silvery grey appressed-woolly leaves. It is very common in tall alpine herbfield.

340–1 *Craspedia* sp. D. The densely villous leaves of this species are seen in 341 showing the young leaves. The beautiful globular yellow heads are seen in the plant growing on the slopes above Blue Lake.
342 *Craspedia* sp. E is a common species with less hairy leaves than sp. D and growing here near Lake Cootapatamba.

343 *Craspedia* sp. F has pale green leaves with dense short rather stiff glandular hairs and usually orange flower heads. It is mostly found in drier sites near rocky outcrops or on shallow rocky soil. It is seen here surrounded by the very common introduced *Rumex acetosella* (Sheep Sorrel).

341

343

342

344 *Cotula alpina* (Alpine Cotula) is a small stoloniferous herb often forming mats, recognizable by its
fleshy pinnatifid leaves and small congested flower
heads on thick stalks which elongate in fruit.

345 *Abrotanella nivigena* (Snow-wort) is a dwarf mat-forming herb characterized by its crowded thick shiny leaves and short-stalked flower heads. It is readily distinguished from the other dwarf alpine daisy *Parantennaria uniceps* by the obtuse, not sharp-pointed leaves, the thick phyllaries and the absence of pappus hairs on the fruit.

346 *Senecio gunnii* differs from the other two species of *Senecio* in its less conspicuous flower heads which lack the large petal-like ligules.

345

346

347 *Senecio gunnii*.

348 In *Senecio pectinatus* (Alpine Groundsel) the individual flower heads are large and occur singly on the reddish stalks. The thick, slightly fleshy leaves form basal rosettes.

349 *Senecio lautus* subsp. *alpinus* (Variable Groundsel) is a common species with deeply divided leaves and usually numerous yellow flower heads.

347

348

349

350 *Microseris lanceolata* (Native Dandelion) has large yellow flower heads on long stalks and drooping buds. *Microseris* tends to flower later than most other alpine species and is conspicuous in the tall alpine herbfields and sod tussock grasslands during February and March.

350

leaves mostly basal, the petioles stem-clasping at the base, the blade once or twice pinnatisect, the primary segments up to 1 cm long, at least the distal segments toothed or again divided into linear lobes ± 1–2 mm wide; *flowering stems* with a few cauline leaves at the base grading upwards to small linear entire bracts, sometimes only a few linear bracts present or the stems ebracteate; *capitula* (2–)3–4.5 cm diam. including the white ligules of the ray florets; *phyllaries* narrow-oblong to oblong-lanceolate, blunt to subacute, minutely lacerate distally; *receptacle* conical at maturity; *achenes* 2–3 × 1–2.5 mm, flat with minute glandular hairs up the centre of the oblong-cuneate body and on the margins of the broad entire or shortly lobed wings; *pappus* conspicuous; $n = 11^*$.

Alpine and subalpine tracts of the Australian Alps and adjacent ranges from the summit of Mt Gingera in the A.C.T. to Mt Wellington in Victoria. *Distribution*

One of the most spectacular of the mountain daisies, it is a characteristic component of *Brachycome–Danthonia* tall alpine herbfield, and is occasionally found in *Celmisia–Poa* tall alpine herbfield. *Notes and Habitat*

References: Davis (1948); Smith-White *et al.* (1970)*.

B. nivalis F. Muell. var. *alpina* (F. Muell. ex Benth.) G. L. Davis, Proc. Linn. Soc. N.S.W. 73: 198 (1948)

Stoloniferous perennial herb to about 20 cm high from branched ascending rhizomes; *leaves* to ± 5 cm long, linear to narrowly obovate-spathulate, obtuse to subacute, expanded and stem-clasping at the base, entire or irregularly toothed, or pinnatipartite with narrow linear lobes to 4.5 mm long, sometimes grading into linear bracts up the stem, or the stem bracts few or rarely absent; *flowering stems* ± densely puberulent with tiny glandular-septate hairs distally; *capitula* 2.5–4 cm diam., the ligules of the ray florets white; *achenes* similar to those of the type variety; $n = 9 + 0–2B^*$. [299]

Alpine and subalpine tracts of the Australian Alps. *Distribution*

Wet areas in short alpine herbfield, wet depressions in sod tussock grassland and tall alpine herbfields, and on stream banks. Occasional specimens of this variety with entire leaves and bracts may be difficult to distinguish from some forms of *B. obovata* in the absence of fruit; however, the flowering stems of the latter species are almost glabrous. Depauperate specimens with entire leaves and bracts might also be confused vegetatively with robust plants of *B. stolonifera* (see Notes under that species). Smith-White *et al.* (1970) suggest that this variety merits specific rank. *Notes and Habitat*

References: Davis (1948); Smith-White *et al.* (1970)*.

B. tenuiscapa Hook. f., Hook. Lond. J. Bot. 6: 114 (1847)
var. *tenuiscapa*

Stoloniferous perennial herb (5–)10–20 cm high with glandular-septate indumentum, often forming extensive patches from the strongly rooting stolons which are 3–5 mm in diameter; *leaves* ± 2–2.5 × 0.5–1 cm, obovate-spathulate, [300] the blade crenate-serrate, the broad pseudo-petiole slightly stem-clasping at the base, mostly radical but grading into 4–13 bracts up the stem; *capitula* 2.5–3 cm diam., including the pale violet or almost white ligules of the ray florets; *phyllaries* 16–22, oblanceolate, obtuse to subacute, glandular on the back; *achenes* black at maturity, glabrous, 2–2.4 × 1–1.3 mm, obovate to obovate-cuneate with a narrow, slightly thickened margin; *pappus* minute.

Distribution   Mt Kosciusko area, N.S.W., the Bogong High Plains, Victoria, and mountains of Tasmania.

Notes and Habitat   The above description applies to the mainland populations of this variety, which were described as *B. alpina* P. F. Morris (1924) non Colenso (1898). These seem to be generally more robust than Tasmanian specimens, with more numerous stem bracts and broader, more evenly toothed leaves. Colonizing bare patches in sod tussock grassland and *Celmisia–Poa* tall alpine herbfield.

References: Davis (1948); Morris (1924).

B. stolonifera G. L. Davis, Proc. Linn. Soc. N.S.W. 74: 145 (1949)

[301]   Small glabrous stoloniferous perennial herb about 3–7 cm high; *stolons* ± 5–10 cm long; *leaves* radical, entire, ± 1–3 × 0.1–0.2 cm, linear or narrow oblong, obtuse, stem-clasping at the base; *flowering stems* simple, ebracteate or with a small linear bract; *capitula* solitary, 1.5–3 cm diam. including the white ligules of the ray florets; *phyllaries* ± 4.5 × 1.4 mm, obovate to oblanceolate, blunt to subacute, minutely lacerate at the apex and usually fringed with microscopic glandular-septate hairs, green becoming ± stained with reddish purple; *achenes* obovate-cuneate, 1.8–2.1 × 0.9–1.2 mm, red-brown, the thick central region and the edge of the narrow thin margin covered with minute tubercle-based glandular-septate hairs; *pappus* of conspicuous silky white bristles; $n = 15^{\star}$.

Distribution   Endemic to the alpine tract of the Kosciusko area, N.S.W.

Notes and Habitat   Occasional depauperate plants of *B. nivalis* var. *alpina* with entire leaves and bracts can be confused with *B. stolonifera*, but these can usually be distinguished by their more numerous stem bracts. Fairly common in short alpine herbfield and wet sites in tall alpine herbfield and sod tussock grassland.

References: Davis (1948); Smith-White *et al.* (1970)*.

B. scapigera (Sieb. ex Spreng.) DC., Prodr. 7: 277 (1838)
'Tufted daisy'

[302]   Erect glabrous ± densely tufted perennial herb ± 12–30(–40) cm high, the base of the plant surrounded by the fibrous remains of old leaves; *leaves* radical, entire, (3–)6–15(–19) cm × 5–15 mm, narrowly to broadly oblanceolate, attenuate to the base, the midrib prominent on the underside; *flowering stem* ebracteate or with 1–2 small linear bracts; *capitula* 1.5–2.5(–3) cm diam. including the white or pale violet ligules of the ray florets; *phyllaries* 12–15(–18), obovate to broadly elliptical or oblanceolate, blunt or obtuse, rarely subacute; *fruiting receptacle* ± swollen; *achenes* glabrous, obovate to obovate-cuneate, ± 2–3 ×

0.8–1.2 mm, flattened, with very narrow slightly thickened smooth margins; *pappus* minute and inconspicuous.

South-eastern Queensland (Stanthorpe area), tablelands, Dividing Range and south coast of N.S.W., and highlands of Victoria to the Grampians. *Distribution*

Common in several communities including tall alpine herbfields, sod tussock grasslands, heaths and margins of bogs. See Notes under *B. obovata* for differences from that species. *Notes and Habitat*

Reference: Davis (1948).

B. obovata G. L. Davis, Proc. Linn. Soc. N.S.W. 74: 146 (1949)

Glabrous perennial herb 10–20(–30) cm high, loosely tufted or shortly creeping at the base from ascending rhizomes; *leaves* entire, mostly radical, or cauline leaves developed on ascending stems in wet situations, ± 3–15 cm × 1.1–6 mm, linear to narrowly oblanceolate, usually long-tapering proximally, slightly expanded and stem-clasping at the base; *flowering stems* glabrous or with very sparse minute glandular hairs, mostly ± 4–6 small linear entire bracts; *capitula* 2–3 cm diam., the ligules of the ray florets white or faintly tinged with mauve; *phyllaries* 20–29, acute or subacute, usually serrulate distally; *achenes* smooth, obovate, ± 2.3–2.5 × 1–1.3 mm, somewhat reddish at maturity; *pappus* straw-coloured, small and inconspicuous. [303]

High elevations in the Australian Alps. *Distribution*

Wet areas on the margins of fens and bogs, and stream banks. Some forms of this species bear a close vegetative resemblance to forms of *B. radicans*, which occurs at lower altitudes, and to forms of *B. nivalis* var. *alpina* with entire leaves and bracts (see Notes under that species). *B. obovata* may also resemble narrow-leafed forms of *B. scapigera*, but may be distinguished from the latter by the less densely tufted habit, the usually more numerous stem bracts, the more numerous acute or subacute phyllaries, the slightly more conspicuous pappus, and the lack of fibrous remains of old leaves at the base of the plant. *Notes and Habitat*

Erigeron L.

E. setosus (Benth.) M. Gray, Contrib. Herb. Aust. No. 6: 1 (1974)

Small perennial herb from much-branched ascending or creeping rhizomes, forming dense patches up to 30 cm or more in diam.; *rhizomes* 1–3 mm diam.; *scapes* up to 7 cm high in fruit, hispid with septate hairs and also with dense [304–6] microscopic glandular hairs, usually with a linear bract at or below the middle; *leaves* crowded in rosettes ± 1.5–3.5 cm diam., petiolate or subsessile, ± 0.7–1.5 cm long, obovate to spathulate, entire, coarsely setose with septate hairs, the midrib prominent on the underside; *petiole* stem-clasping at the base, almost glabrous except the margins fringed with septate hairs; *capitula* solitary, 0.5–1 cm diam., sessile or almost so in flower, the fruiting heads globular, ± 1.5 cm diam. and raised on scapes ± 4–7 cm high; *phyllaries* linear, 5–6 mm long, acute or acuminate, with sparse septate hairs and dense microscopic glandular hairs, usually tinged with reddish purple distally; *ligules* of ray florets white, 1–2 mm long; *achenes* glabrous, ± 2.5 mm long, narrowly obovate-cuneate; *pappus bristles* 3.5–4 mm long.

Distribution   Endemic to the alpine tract of the Kosciusko area, N.S.W.

Notes and   A very distinctive little species, fairly common in short alpine herbfield and
Habitat   on wet areas along stream banks.

E. pappocromus Labill., Nov. Holl. Plant. Specim. 2: 47, t. 193 (1806), sens. lat.

The taxonomy of this polymorphic species has not been fully worked out (see Curtis 1963); the following forms occur in the Kosciusko area:

---

1   Involucre ± 1.3 cm long; leaves mostly 1–2 cm wide; rootstock branched, shortly creeping, ± woody, ± 3–4 mm diam.   2
1   Involucre ± 8 mm long; leaves pilose, 0.3–1 cm wide; smaller and more slender than forms *B* and *C*, from slender long-creeping rhizomes 1–2 mm diam. and short stolons; alpine and subalpine tracts of the Kosciusko area, and the Bogong High Plains, Victoria, often growing through *Sphagnum*
    form A

---

[307]   2   Leaves ± scabrid and viscid; widespread and fairly common above the treeline in alpine and subalpine tracts of the Kosciusko area, and the Bogong High Plains, Victoria   *form B*
2   Close to the preceding, differing mainly in the pubescent, non-viscid leaves; rather uncommon above the treeline, but common in the subalpine tract; also in the Victorian Alps   *form C*

Celmisia Cass.

[308–11]   The Kosciusko species of *Celmisia*, collectively known as Silver Snow Daisies, have been referred to *C. longifolia* Cass. in the literature. Given (1969), however, has shown that Bentham's circumscription of that species was too wide, and is at present revising the Australian species. The common *Celmisia* above the treeline, which is a characteristic species of the *Celmisia–Poa* tall alpine herbfield, is one of the most important and abundant plants in the alpine flora. It forms

extensive, almost pure communities, which give a distinctive silvery grey patch-work appearance to the alpine landscape, changing to white when the mass flowering occurs.
Reference: Given (1969).

Olearia Moench

O. algida N. Wakef., Victorian Nat. 73: 97 (1956)
'Alpine daisy-bush'

Shrub about 0.5–1 m high; *branchlets* thinly cottony-tomentose with short white crisped hairs and minute ± hidden glandular hairs, soon glabrescent; *leaves* very small, sessile, ± 1.5–2.5 mm long, clustered on tiny short shoots along the branchlets, thick, blunt, narrowly ovate-triangular in outline, expanded and auriculate at the base, glabrous and shiny above, the white-tomentose underside almost hidden by the revolute margins; *capitula* sessile and terminal on short leafy branchlets; *involucre* ± 4–4.5 mm long; *phyllaries* pale green, sometimes pink-tinged, obtuse, glabrous or almost so, with narrow hyaline ciliolate margins; *female florets* 3–5, with white ligules ± 2–3.5 mm long; *disc florets* 3–5; *achenes* glabrous or puberulent; *pappus bristles* ± 3.5 mm long.   [312]

Alpine and subalpine tracts of the Australian Alps, and mountains of Tasmania.   *Distribution*

Found in heaths not far beyond the treeline. The leaves of this species are very similar in shape to those of *Helichrysum hookeri* but are not closely appressed to the stem as they are in that species.   *Notes and Habitat*

O. phlogopappa (Labill.) DC. var. *subrepanda* (DC.) J. H. Willis, Muelleria 1: 32 (1956)

Small greyish shrub 30–75 cm high with tomentose branchlets; *leaves* small and crowded, sessile and slightly stem-clasping at the base, 5–10(–15) × 2–6 mm, obovate-oblong, obtuse or rounded at the apex, obscurely repand-denticulate or subentire, greyish and stellate-hairy above, densely yellowish stellate-tomentose on the underside becoming grey with age; *capitula* ± 1.5–2 cm diam., usually   [313] solitary and terminal on short leafy lateral shoots, the peduncles not or only shortly exceeding the uppermost leaves; *involucre* 5–8 mm long; *phyllaries* in several rows, herbaceous, ± tomentose on the back, with narrow scarious ciliate and usually purple-tinged margins; *ray florets* 14–20, female, with white ligules; *disc florets* tubular; *achenes* 1.5–2.5 mm long, antrorsely pubescent with simple or forked hairs and usually sparse microscopic sessile glands; *pappus bristles* barbellate, 5–7 mm long, with an outer row of short ± flattened laciniate bristles about 1 mm long.

Distribution   Alpine and subalpine tracts of the Australian Alps and associated ranges as far north as the A.C.T., and mountains of Tasmania.

Notes etc.   Common in heaths, typically in rocky sites.

O. phlogopappa (Labill.) DC. var. *flavescens* (Hutch.) J. H. Willis, Muelleria 1: 32 (1956)

[314]   Close to the preceding, differing mainly in the larger leaves which are mostly 20–50 × 5–10(–15) mm, and the flowers distinctly overtopping the leaves on long bracteate peduncles.

Distribution   Alpine and subalpine tracts of the Australian Alps.

Notes and Habitat   Fairly common in heaths and large granite outcrops. In the subalpine form of this variety the leaves are usually longer and shortly petiolate.

Gnaphalium L.

KEY TO THE SPECIES

1   Innermost phyllaries deeply laciniate, ± 4 mm long; capitula clustered, subtended by 1–3 broad spreading elliptical or obovate leafy bracts; rare plant of rocky cliffs or outcrops near water   *G. umbricola* (p. 367)
1   Innermost phyllaries not laciniate; capitula, if clustered, subtended by oblanceolate to linear bracts   2

2   Involucre ± 5–6 mm long; inner phyllaries 4–5 mm long; pappus bristles 3–4 mm long   *G. argentifolium* (p. 367)
2   Involucre ± 7 mm long or more; inner phyllaries ± 6 mm long or more; pappus bristles 5 mm long or more   3

3   Capitula solitary, the involucre ± 9–10 mm long, subsessile or raised on weak peduncles up to 3 cm long at maturity; leaf blades ± 5–7 × 2–3 mm   *G. nitidulum* (p. 368)
3   Capitula numerous, the involucre ± 7–8 mm long, the flowering stems ± 5–15 cm long at maturity; leaf blades ± 1.5–3 × 0.5–0.8 cm   *G. fordianum* (p. 368)

G. umbricola J. H. Willis, Victorian Nat. 73: 200 (1957)
'Cliff cudweed'

Tufted perennial herb about 5–15 cm high, sometimes forming leafy patches [315] by means of stolons; *flowering stems* weak, erect or ascending, densely covered with appressed woolly hairs; *rosette leaves* long-petiolate, the petioles expanded with glabrescent margins at the base, the blades obtuse, ± 1–6 × 0.5–1.8 cm, narrowly or broadly elliptical, oblanceolate or obovate, the upper surface greyish with appressed white webby hairs or green and glabrescent, densely white-tomentose on the underside; *cauline leaves* similar to the rosette leaves but smaller; *capitula* up to ± 7, in a dense spherical terminal cluster subtended by a few conspicuous spreading leafy bracts which resemble the upper cauline leaves, sometimes also a few smaller clusters of capitula in the upper axils; *involucre* ± 4.5 mm long, woolly at the base, the innermost phyllaries deeply laciniate at the apex; *achenes* slightly flattened or almost subangular in cross-section, ± 1–1.4 × 0.3–0.5 mm, ± sparsely covered with microscopic antrorsely appressed finger-like hairs; *pappus bristles* ± 2.5 mm long, free at the base.

Mountainous areas of south-eastern Australia as far north as Fitzroy Falls, *Distribution* N.S.W., and Tasmania.

As Willis (1957) points out, this uncommon species is usually found in the shade, *Notes and* perching on wet rock faces or ledges and often associated with waterfalls or *Habitat* cascades. It is rare above the treeline, being found in crevices in rocky crags above Blue Lake and Lake Albina, in *Brachycome–Danthonia* tall alpine herbfield. It is more common in the subalpine tract. Because of its very specialized habitat this species is rather sparsely distributed and the individual populations are usually very small, often consisting of only a few plants. Where not strictly protected it could be vulnerable to over-collecting. Drury (1972) refers to this species as being subdioecious, there being considerable variation in the proportion of female to disc florets in the capitula.

References: Curtis (1963); Drury (1972); Willis (1957).

G. argentifolium N. Wakef., Victorian Nat. 73: 187 (1957)
'Silver cudweed'

Low mat-forming perennial from thin, wiry, ascending or shortly creeping [316] branched rhizomes; *flowering stems* 3–8(–15) cm high, weak, ascending, loosely white-cottony, with the dried brown remains of old leaves at the base; *leaves* silvery or whitish, mostly basal, petiolate, oblanceolate to obovate-spathulate, densely woolly on both sides except for the small glabrous mucronate tip; *capitula* ovoid or hemispherical, solitary or few together and almost sessile, or up to ± 7 on long peduncles, the involucre 5–6 mm long; *outer phyllaries* ovate, scarious, glabrous except with long woolly hairs at the base; *inner phyllaries* 4–5 mm long, the stereome pale yellowish green, the lamina scarious, very pale stramineous, brownish towards the base, tinged with red near the apex of the stereome when young; *achenes* 0.8–1.1 × 0.3–0.4 mm, slightly flattened, densely micropapillate and with sparse microscopic antrorsely appressed finger-like hairs; *pappus bristles* 3–4 mm long, weakly adhering and ciliolate at the base.

Distribution   Alpine and subalpine tracts of the Australian Alps.

Notes etc.   Fairly common in tall alpine herbfields and sod tussock grassland.

G. nitidulum Hook. f., Handb. N.Z. Fl.: 154 (1864)
'Shining cudweed'

[317]   Low cushion- or mat-forming perennial, growing in patches up to 15 cm or more in diam. from a branched ascending or shortly creeping rootstock; *aerial stems* numerous, crowded, only a few cm high in the vegetative state, densely covered at the base with the dried remains of old leaves; *leaves* crowded, with spreading oblong-spathulate blades ± 5–7 × 2–3 mm, slightly keeled distally, blunt, densely covered with stiff appressed shining silvery hairs on both surfaces, becoming yellowish grey or ferruginous with age; *petioles* broad, flat, appressed and closely imbricate, about one-third the length of the leaf, ± glabrous adaxially, woolly on the back and margins distally; *capitula* solitary, terminal, sessile or subsessile, the involucre ± 9–10 × 5–7 mm in flower, in the fruiting stage subsessile or raised on a rather weak loosely cottony peduncle up to 3 cm long with one to several leafy bracts; *phyllaries* glabrous and shining except the lowest loosely woolly at the base, the inner ± 7–8.5 mm long, the stereome yellowish green, the scarious lamina very pale stramineous becoming brownish towards the base; *achenes* slightly flattened, ± 1.3–1.5 × 0.4–0.5 mm, rather obscurely micropapillate; *pappus bristles* ± 6–6.5 mm long, weakly adhering at the base.

Distribution   Alpine and subalpine tracts of the Kosciusko area, N.S.W., Pretty Valley (Bogong High Plains), Victoria, and New Zealand (South Island).

Notes and Habitat   Wet areas near streams and near the margins of bogs, and sod tussock grassland. This species could be confused with small plants of *Ewartia nubigena* in the vegetative state.

Reference: Drury (1972).

G. fordianum M. Gray, Contrib. Herb. Aust. No. 26: 2 (1976)

[318]   Perennial herb, loosely tufted or more frequently creeping and rooting at the base and forming leafy patches up to 15 cm or more in diam.; *flowering stems* lanuginose, leafy, erect or ascending, ± 3–15 cm high, often elongating and becoming rather weak at maturity, with dried brown remains of old leaves at the base; *leaves* narrowly oblanceolate to obovate-spathulate, the blade ±1.5–3 × 0.5–1 cm, silvery-pannose, the midrib conspicuous on the underside, the apex with a glabrous, often curved, mucronulate tip; *petiole* shorter than the blade, slightly expanded and glabrescent towards the base; *capitula* ± 5–15, ovoid-cylindrical, at first congested in a terminal bracteate cluster, the branches and peduncles usually elongating at maturity to form a ± open paniculate inflorescence; *involucre* ± 7–8 mm long, woolly at the base; *phyllaries* shining, the outer ovate with woolly hairs on the short stereome at the base, the inner glabrous, narrowly elliptic to narrowly oblanceolate, obtuse or subacute, ± 6–7 mm long, the lamina stramineous at maturity, ± tinged with red around the apex

of the greenish stereome; *hermaphrodite florets* ± 3–13, with ± 35–56 female florets per capitulum; *achenes* 1.2–1.4 × 0.3–0.4 mm, slightly flattened, densely micropapillate; *pappus bristles* 5–5.5 mm long, connate in a ring and ciliolate at the base.

Alpine and subalpine tracts of the Kosciusko area, N.S.W. as far north as Mt Jagungal, the Bogong High Plains and Lake Mountain in Victoria, and Pine Lake, Central Plateau, Tasmania.

Distribution

Rather sparsely distributed, colonizing bare patches in tall alpine herbfield and sod tussock grassland, usually on slopes near rivers and creeks.

Notes and Habitat

Ewartia Beauverd

E. nubigena (F. Muell.) Beauverd, Bull. Soc. Bot. Genève (2)2: 239 (1910)   'Silver ewartia'

Functionally dioecious, silvery-grey mat-forming perennial subshrub; *stems* much branched, elongate, ± 2 mm diam., creeping and rooting or sprawling over rocks, ± covered with the remains of old leaves, the branchlets ascending and 3–6(–10) cm high; *leaves* crowded, ±4–7 × 2 mm, obovate-oblanceolate to subspathulate, ± recurved towards the apex, with dense appressed silvery grey tomentum becoming brownish with age, the broad imbricate bases glabrous adaxially; *capitula* sessile, functionally unisexual, 5–8 mm diam., terminal and usually solitary on the branchlets; *outer phyllaries* scarious with inconspicuous whitish tips and webby hairs towards the base; *inner phyllaries* glabrous, with conspicuous white oblong radiating laminae; *male capitula* consisting entirely of hermaphrodite florets which very rarely set seed; *female capitula* with several rows of fertile female florets with filiform corollas irregularly 3–4-toothed at the apex, the disc florets 5-toothed, with sterile stamens, usually setting seed; *achenes* 1–1.3 × ± 0.5 mm, subterete, appressed-puberulent; *pappus bristles* caducous ± 4 mm long, barbellate, ciliolate and weakly cohering at the base, thickened towards the tips except in the female florets.

[319–20]

Alpine and subalpine tracts of the Australian Alps.

Distribution

This species sometimes forms extensive mats or hummocks which are frequently almost covered by the abundant small 'everlasting' capitula. It is an important component of *Epacris–Chionohebe* feldmark and commonly colonizes bare areas in tall alpine herbfields.

Notes and Habitat

Parantennaria Beauverd

P. uniceps (F. Muell.) Beauverd, Bull. Soc. Bot. Genève (2) 3: 256 (1911)

[321–2] Dwarf prostrate dioecious perennial herb, usually forming small patches, cushions or short dense swards; *stems* branched, creeping and rooting, ± 1–1.5 mm diam.; *leaves* glabrous, crowded, linear 5–10(–15) × 0.5–1 mm, stiff and subcoriaceous, mucronate, the broad sheathing bases fringed with woolly hairs and often tinged with reddish purple; *capitula* solitary, terminal, the involucre ± 5 mm long; *phyllaries* scarious, pale brown to stramineous, tinged with reddish purple near the apex of the stereome, the outer ovate, glabrous except for sparse woolly hairs at the base, the inner glabrous, oblong-lanceolate to linear-lanceolate with acute or acuminate tips; *male capitula* sessile, hemispherical, the florets ± 4 mm long with 5(–6)-toothed corollas, the stamens partially exserted, the styles usually undivided; *female capitula* at first sessile, ovoid-cylindrical and narrower than the male, soon expanding and raised on weak sparsely woolly bracteate peduncles 1–2 cm long at maturity, the corollas 4-toothed and filiform, the styles 2-fid; *achenes* glabrous, terete, ± 1 × 0.4 mm; *pappus bristles* barbellate ± 3.5–4.5 mm long.

Distribution Alpine and subalpine tracts of the Kosciusko area, N.S.W., rare on Mt Gingera, A.C.T., and near Mt Cope (Bogong High Plains), Victoria.

Notes and Habitat This monotypic genus is endemic to the high country of south-eastern Australia; it is found mainly in short alpine herbfield and also in wet areas in sod tussock grassland, and on the margins of streams and bogs.

Helipterum DC.

KEY TO THE SPECIES

1 Decumbent herb with broad, densely lanuginose obovate to oblanceolate leaves; achenes glabrous   *H. albicans* subsp. *alpinum* (p. 370)

1 Erect tufted herb with linear subglabrous leaves which are conspicuously glandular-pitted (under 10× lens); achenes silky-hairy
                                                              H. anthemoides (p. 371)

H. albicans (A. Cunn.) DC. subsp. *alpinum* (F. Muell.) P. G. Wilson, Trans. R. Soc. S. Aust. 83: 174 (1960)   'Alpine sunray'

[323–5] Decumbent perennial herb 10–20(–24) cm high from branched woody creeping rootstock, often forming extensive patches up to several metres in diameter; young stems densely lanuginose, covered at the base with the remains of old leaves; *leaves* soft, thick, congested, densely lanuginose, 1.5–5 × 0.5–1 cm, obovate, oblanceolate or obovate-spathulate, obtuse or rounded at the apex, the margins ± recurved, the midrib prominent on the underside, decreasing in size up the stem and merging with the linear bracts of the peduncle; *peduncle* terminal, lanuginose, 2–15 cm long; *capitula* solitary, 2.5–4 cm diam.; *outer phyllaries* sessile, with scarious, shining, pale or stramineous ovate to lanceolate laminae, sometimes variously tinged with pale brown or reddish purple, woolly at the base; *intermediate phyllaries* similar but with slender woolly claws; *inner phyllaries* with milky white spreading petaloid laminae, often with a brown or purplish spot at the base; *disc* yellow-orange; *achenes* glabrous, ± 1 mm long, obovoid, obscurely 4-angled, slightly curved; *pappus bristles* plumose, 4.5–5.5 mm long.

Alpine and subalpine tracts of the Australian Alps and associated ranges, as far north as the Brindabella Range, A.C.T. *Distribution*

This strongly perfumed everlasting, with its masses of orange and white papery capitula and silvery leaves, is common in exposed sites in tall alpine herbfields and in *Epacris–Chionohebe* feldmark. Wilson (1960) points out that this alpine subspecies grades into the forma *purpureo-album* of subsp. *albicans* at lower elevations on the northern and southern limits of its range. *Notes and Habitat*

Reference: Wilson (1960).

H. anthemoides (Sieb. ex Spreng.) DC., Prodr. 6: 216 (1838)
'Chamomile sunray'

Tufted perennial herb 15–30(–40) cm high from a tough erect woody rootstock; *stems* numerous, erect, slender and wiry, usually simple, almost glabrous below, glandular-pubescent and sparsely woolly-hairy below the capitula; *leaves* cauline, crowded, suberect, ± 10–20 × 0.1–0.3 mm, conspicuously glandular-pitted, obscurely serrulate and with sparse minute septate hairs on the margins; *capitula* solitary, terminal, 2–2.5(–3) cm diam.; *outer phyllaries* broadly elliptic, scarious, shining, with dark midnerve and hyaline fimbriate margins; *inner phyllaries* with conspicuous petaloid milky white radiating laminae 6–10 mm long, often with a dark spot at the base, the florets all tubular, hermaphrodite; *achenes* densely silky-hairy, 2–3.5 × ± 1.5 mm, obovoid, slightly flattened; *pappus bristles* plumose, ± 4 mm long, slightly flattened towards the base. [326]

Widespread in eastern Australia and Tasmania. *Distribution*

This white-rayed everlasting, although widespread at lower altitudes, is uncommon in the alpine tract as small colonies in tall alpine herbfield. *Notes and Habitat*

Helichrysum Mill.

1 Herb with large solitary capitula, the petaloid tips of the inner phyllaries yellow   *H. scorpioides* (p. 372) KEY TO THE SPECIES
1 Shrubs with small clustered capitula, the petaloid tips of the inner phyllaries milky-white   2

2	Leaves erect and closely appressed to the stem, 1–3 mm long, triangular-ovate with expanded appressed auriculate bases *H. hookeri* (p. 372)
2	Leaves spreading, 5–10 mm long, not as above 3

3	Capitula with ± 15–20 florets, about as wide as or wider than long at maturity; leaves with a pungent spicy aroma, white- or greyish-tomentose on the underside *H. secundiflorum* (p.373)
3	Capitula with ± 3–6 florets, longer than wide; leaves not aromatic, golden-tomentose on the underside when young *H. alpinum* (p. 373)

H. scorpioides Labill., Nov. Holl. Plant. Specim. 2: 45, t. 191 (1806)
'Button everlasting'

[327]   Perennial herb 15–25(–30) cm high from slender branched rhizomes; *stems* usually simple, decumbent or suberect, loosely woolly; *leaves* soft, sessile, obovate to oblanceolate, 2.5–4 × 0.4–0.6 cm, becoming smaller and grading into bracts up the stem, woolly on both sides or glabrescent above, with a small glabrous mucro at the apex; *capitula* solitary and terminal, ± 2–3 cm diam.; *outer phyllaries* scarious, tinged with pale brown, conspicuously woolly on the margins towards the base, often ± wrinkled and lacerate distally; *inner phyllaries* with green ± woolly claws and ± spreading yellow petaloid laminae which are sometimes wrinkled and orange-tinged on the outside; *florets* very numerous, the outer rows pistillate; *achenes* glabrous; *pappus bristles* weak, minutely barbellate, 5–7 mm long, slightly thickened and golden-yellow distally.

Distribution   Widespread in south-eastern Australia and Tasmania.

Notes and Habitat   As Willis (1973) points out, this variable species forms a complex with *H. rutidolepis* DC. and needs revision; the above description applies to the high-altitude form of the complex. It is locally common but very sparsely distributed above the treeline, in tall alpine herbfield.

Reference: Willis (1973).

H. hookeri (Sond.) Druce, Rep. Bot. (Soc.) Exch. Cl. Manchr 1916: 626 (1917)
'Scaly everlasting; kerosene bush'

[328]   Aromatic shrub about 0.5–1 m high; *branchlets* densely white-tomentose; *leaves* sessile, viscid, scale-like, crowded, ± 1–3 mm long, erect and closely appressed to the stem, triangular-ovate with expanded and appressed auriculate bases, the revolute margins stuck together with resin or parting slightly to reveal the white-tomentose undersides; *capitula* cylindrical, ± 4–5 mm long, sessile in small hemispherical or subglobose clusters about 1–1.5 cm diam. at the ends of the upper branchlets; *phyllaries* slightly viscid, the outer scarious, yellowish green or stramineous with indurated centres and narrow hyaline margins, the inner with small milky-white radiating tips ± 1 × 0.8 mm; *florets* 2–4; *achenes* 1–1.2 mm long, minutely papillose; *pappus bristles* 3–3.5 mm long, slightly thickened distally.

Alpine and subalpine tracts of the Australian Alps and associated ranges as far north as the Brindabella Range, A.C.T., and high mountains of Tasmania. *Distribution*

Whereas this species is very common in the subalpine tract in bogs and heaths, it tends to be rather marginal above the treeline. As the common name Kerosene Bush indicates, it burns readily; however, it is adapted to fires and after burning usually recovers more densely than before. *Notes and Habitat*

Reference: Burbidge (1958).

H. secundiflorum N. Wakef., Victorian Nat. 68: 49 (1951)
'Cascade everlasting'

Aromatic shrub about 0.5–1.5 m high (to 2 m high with long floriferous spreading branches at lower altitudes); *branchlets* densely white- or greyish-tomentose; *leaves* narrowly oblong or narrowly oblanceolate, 6–10(–15) × 1.5–2(–3) mm, viscid and with ± pinnately arranged appressed webby hairs above, densely greyish-tomentose on the underside; *capitula* campanulate, 3.5–4 × 5–6 mm, in small rounded terminal clusters which are usually crowded and secund along the main stems on short lateral branches; *outer phyllaries* scarious, shining, pale yellowish brown or tinged with pale pink, the inner with conspicuous milky-white spreading laminae; *florets* 15–20, a few of the outer row female and filiform; *achenes* puberulent; *pappus bristles* 3–3.5 mm long, very slightly thickened towards the apices. [329]

Mainly found in the alpine and subalpine tracts of the Australian Alps and associated ranges as far north as the A.C.T. *Distribution*

This shrub is often very spectacular in the subalpine tract, with long arching densely floriferous branches, with respect to which the common name Cascade Everlasting is most appropriate. However, above the treeline where it occasionally occurs in heaths, it is a smaller shrub usually found in the shelter of large boulders or other heathland shrubs. The leaves have a very distinctive pungent spicy aroma. *Notes and Habitat*

Reference: Burbidge (1958).

H. alpinum N. Wakef., Victorian Nat. 68: 49 (1951)
'Alpine everlasting'

Shrub about 0.75–1(–1.5) m high, the branchlets and undersides of the leaves with dense golden-yellow tomentum becoming greyish with age; *leaves* sessile, crowded, widely spreading, oblong, obtuse, 4–10 × 2–3 mm with tightly recurved margins, the upper surface greyish green and ± viscid with pinnately-arranged appressed webby hairs or glabrescent, the underside densely yellow-tomentose with prominent midrib; *capitula* cylindrical or narrow turbinate, very shortly stalked, 4.5–7 mm long, arranged in dense rounded terminal clusters; *outer phyllaries* pale brown, usually strongly tinged with red, with sparse webby hairs on the back and hyaline apices; *inner phyllaries* with conspicuous [330] blunt milky-white radiating petaloid laminae ± 1–2 mm long; *florets* 3–6; *achenes* 1–1.3 mm long, microscopically appressed-puberulent; *pappus bristles* barbellate, 3.5–4 mm long, not or scarcely thickened at the apex.

Distribution   Alpine and subalpine tracts of the Australian Alps.

Notes etc.   Fairly common in bogs and heaths.

Reference: Burbidge (1958).

Leptorhynchos Less.

L. squamatus (Labill.) Less., Syn. Gen. Compos. 273 (1832)
'Scaly buttons'

[331–2]   Very variable perennial herb 5–25(–30) cm high; *stems* leafy, ascending or decumbent at the base, loosely woolly; *leaves* sessile, ± 1.5–3.5 cm long, narrowly oblanceolate, elliptic or oblong, acute with a short glabrous mucro, variously hairy or green and glabrescent above, densely appressed-woolly on the underside, the margins recurved; *capitula* 10–15 mm diam., campanulate with the outer florets curved and overhanging the involucre, solitary on a long slender wiry penduncle with numerous small ± appressed membranous bracts which grade into the phyllaries above; *phyllaries* numerous in many rows, hyaline with pale brown ± recurved tips, densely fringed on the sides with long woolly hairs and with sparse coarse acicular cilia distally, the outer ovate-triangular and sessile, the inner shortly clawed with narrow-elliptic or elliptic-lanceolate acute laminae; *florets* all tubular, a few of the outer florets female, the bisexual florets very numerous, expanded and campanulate above; *achenes* glabrous, terete; *pappus bristles* barbellate, ± 4–5 mm long.

Distribution   Widespread from south-eastern Queensland to Tasmania and South Australia.

Notes and Habitat   A widespread minor herb, forming small colonies in sod tussock grassland and tall alpine herbfields, becoming conspicuous when the bright yellow to orange capitula develop. A small ecotype with the leaves woolly on both sides occurs in the *Epacris–Chionohebe* feldmark. The various forms of this widespread species need revision.

Podolepis Labill.

P. robusta (Maiden & Betche) J. H. Willis, Victorian Nat. 70: 224 (1954)
'Alpine podolepis'

[333–4]   Robust perennial herb 20–40(–60) cm high from a tough woody rootstock; *stems* webby-cottony, densely so above; *radical leaves* ± 8–15(–25) × 2–4.5 cm, moderately thick, bright green and usually glabrous, obovate-spathulate to oblanceolate, rounded to subacute at the apex, tapering to a broad stem-clasping base, the margins ± crinkled; *cauline leaves* becoming progressively smaller up the stem, suberect, narrowly elliptic, acute, stem-clasping and shortly decurrent at the base, the lowest almost glabrous, the upper webby-cottony; *capitula* 3–11 in a terminal subcorymbose cluster, golden yellow, 2–3.5 cm diam.; *involucres* hemispherical, ± 2.5 cm diam.; *outer and intermediate phyllaries* with broadly ovate, obtuse, scarious, shining stramineous laminae, the outer sessile, the intermediate with herbaceous glandular claws which are subequal in length to the laminae; *ray florets* 30–40, the golden-yellow ligules 10–17 mm long, 3–4-toothed distally; *achenes* subterete, 2.5–4 mm long, minutely papillate above and slightly contracted at the apex; *pappus bristles* barbellate, ± 6–8 mm long, united at the base.

Alpine and subalpine tracts of the Australian Alps and associated ranges as far north as the A.C.T.

Distribution

This species is rather uncommon above the treeline in tall alpine herbfield and sod tussock grassland. It is more common in the subalpine tract. Willis (1954) notes: 'Cattlemen sometimes use the name "Mountain Lettuce" for its large lively green rosettes of broad hairless and occasionally crinkled leaves'.

Notes and Habitat

References: Davis (1956); Willis (1954).

Craspedia Forst. f.

As Curtis (1963) and Willis (1973) point out, the *C. glauca* group in Australia is badly in need of revision, and its relationships with the *C. uniflora* group in New Zealand have yet to be worked out. The following taxa can be distinguished from the Kosciusko alpine area, and although some at least are quite well marked and undoubtedly deserve specific rank, others are less well defined and their description is best left until the group as a whole is more adequately understood. For this reason only *C. leucantha* is described below.

PROVISIONAL KEY

Only the most conspicuous types of hairs are referred to in the following key and other types of hairs are usually also present.

1	Florets white or pale cream	2
1	Florets yellow or orange	3

2	Leaves green, glabrous or almost so; margins of wet flushes, springs and small creeks below semi-permanent snow patches; apparently restricted to the alpine tract of the Kosciusko area	
	C. leucantha (p. 376)	
2	Leaves ± silvery-pannose; a small species, usually associated with pools and depressions in sod tussock grassland, wet gravelly areas near the margins of creeks and rivers, and margins of fens; apparently restricted to the alpine tract of the Kosciusko area	[336]
	C. sp. A	

| 3 | Leaves discolorous, the upper surface green or greyish green with dense short glandular-septate hairs and variable rather loose webby hairs, whitish on the underside with closely appressed woolly hairs; florets pale yellow; | [337–8] |

Brachycome–Danthonia tall alpine herbfield and around large rocky outcrops; alpine tract of the Kosciusko area, N.S.W., and Bogong High Plains, Victoria   *C.* sp. B

3   Leaves concolorous   4

[339]   4   Leaves silvery or silvery-grey with appressed woolly hairs on both sides; florets yellow; a large spectacular species, common in *Celmisia–Poa* tall alpine herbfield and drier areas in sod tussock grassland; apparently restricted to the Kosciusko area, mainly in the alpine tract   *C.* sp. C

4   Leaves not as above   5

[340–1]   5   Leaves densely villous with long septate hairs with long filiform apices; florets usually yellow; common in *Celmisia–Poa* tall alpine herbfield and drier areas in sod tussock grassland   *C.* sp. D

5   Leaves not densely villous, but sometimes woolly on the margins   6

[342]   6   Leaves with a ± distinct woolly margin of long hairs with inconspicuous short septate bases; bracts subtending the compound heads not densely glandular, usually broad with conspicuous dark sphacelate scarious margins; florets usually yellow; common in damp areas in tall alpine herbfield and sod tussock grassland   *C.* sp. E

[343]   6   Usually less robust than the preceding, the leaves with dense, short, rather stiff glandular-septate hairs, the long hairs less conspicuous and not or scarcely forming distinct margins; bracts subtending the compound heads densely glandular, usually longer and narrower and with less conspicuous margins than the preceding; florets usually orange (fading to yellow); usually associated with cliffs and rocky outcrops and shallow rocky soils in tall alpine herbfields   *C.* sp. F

Some specimens show characters intermediate between spp. D, E and F, and some degree of introgression may take place between these taxa.

C. leucantha F. Muell., Hook. Kew J. 7: 236 (1855)

[335]   Loosely tufted perennial herb ± 10–50 cm high from a short branched ascending rootstock; *flowering stems* glabrous or with microscopic glandular-septate hairs proximally, loosely cottony above, usually with dried brown remains of old leaves at the base; *basal leaves* green, narrowly oblanceolate to narrowly obovate–spathulate, acute or obtuse with a small blunt callous tip, the blade ± 1–12 × 0.3–2.5 cm, gradually narrowed to a petiole shorter or longer than the blade, appearing glabrous but with variable microscopic glandular-septate hairs and sometimes with inconspicuous cottony hairs on the margins; *cauline leaves* similar to the basal but becoming smaller in size and sessile up the stem to grade into stem-clasping bracts; *compound head* ± 1–2.5 cm diam. at maturity; *bracts* of the general involucre broadly ovate with very broad brown to blackish scarious margins, the stereome ± woolly towards the base; *capitula* 5–8-flowered; *corolla* whitish; *achenes* narrowly turbinate, 3–3.5 × ± 1.2 mm, densely antrorsely hairy; *pappus bristles* plumose, ± 4–5 mm long.

Distribution   Alpine tract of the Kosciusko area, N.S.W.

This species is found on the margins of wet flushes and around springs and small creeks below semi-permanent snow patches and snow drifts. It is markedly variable in size and sometimes flowers in a quite dwarfed state, e.g. about 7 cm high with mature heads about 1 cm diam.
Reference: Gray (1976).

Cotula L.

C. alpina (Hook. f.) Hook. f., Flor. Tas. 1: 192, t. 51A (1856)
'Alpine cotula'

Small stoloniferous creeping perennial herb, apparently glabrous but with minute sparse ± sessile glandular hairs; *leaves* weakly rosulate, petiolate, minutely gland-dotted, the *laminae* 0.5–2 × 0.3–0.6 cm, narrowly elliptic in outline, pinnatifid, the lobes entire or occasionally 1–3-toothed, apiculate; *petioles* sheathing at the base; *capitula* solitary, terminal, 4–7 mm diam., on stout naked erect hollow peduncles which elongate to about 3–5 cm in fruit; *phyllaries* in 2 rows, herbaceous with narrow scarious margins, minutely gland-dotted, rounded and usually reddish-tinged and ± erose at the apex, the outer ovate-oblong, the inner broadly elliptic; *female florets* in several rows, without corollas; *disc florets* functionally staminate with 4-toothed glandular corollas; *achenes* compressed, obovate or obovate-cuneate, 1.7–2 × 1 mm, with a red-brown central portion and pale thick wing-like margins, glandular and occasionally ciliate distally.

Alpine and subalpine tracts of the Australian Alps and associated ranges as far north as the Brindabella Range, A.C.T., and mountains of Tasmania.

Fairly common in the subalpine tract, and occasional in bogs, wet areas in tall alpine herbfields and sod tussock grassland above the treeline.

Abrotanella Cass.

A. nivigena (F. Muell.) F. Muell., Plants Indig. Colon. Vict. 1: text to t. 40 (1865)
'Snow-wort'

Dwarf prostrate mat-forming perennial herb; *stems* numerous, creeping and rooting from branched ascending rhizomes 1–2 mm diam.; *leaves* crowded, the blades (5–)10–15 × 1–1.5 mm, linear-oblong, blunt, glabrous thick and shining, the sheathing bases imbricate with fimbriate margins; *capitula* solitary, terminal, on bracteate peduncles up to 2 cm long; *phyllaries* ± 12–15 in 2–3 rows, 3–3.5 × 0.8–1.5 mm, oblong, blunt, stiff and herbaceous, glabrous except obscurely antrorsely puberulent inside in the lower half, with two pale slightly thickened submarginal nerves; *florets* all tubular, the outer female florets with shortly 2-lobed styles, the disc florets hermaphrodite with styles undivided; *achenes* glabrous, 1.5–1.9 × 0.6–0.8 mm, narrowly obovoid with 3–4 obscure longitudinal ribs; *pappus* absent.

Distribution   Alpine tract of the Kosciusko area, N.S.W., and the Bogong High Plains, Victoria.

Notes and Habitat   This interesting genus of about 20 spp. is mainly subantarctic in distribution, being found in temperate South America, Juan Fernandez and Falkland Islands, New Zealand, Auckland Island, Tasmania, the Australian Alps and high mountains of New Guinea. In the Kosciusko area *A. nivigena* is mostly found in short alpine herbfield and in wet areas near the margins of streams.

Senecio L.

S. gunnii (Hook. f.) Belcher, Ann. Mo. Bot. Gard. 43: 60 (1956)

[346–7]   Perennial herb 20–60(–75) cm high from branched ascending or shortly creeping rootstock; *stems* erect or ascending, often purplish and shortly decumbent at the base, striate, with appressed webby or woolly tomentum, sometimes ± glabrescent; *lower leaves* 5–9(–12) × 1.5–2(–3) cm, obovate to oblanceolate, dentate to subentire, attenuate to the base and subpetiolate, webby-hairy with sepate hairs with long filiform apices, often purplish on the underside, the upper surface sometimes becoming green and glabrescent or hispidulous with persistent hairbases; *upper leaves* smaller with recurved or revolute margins, sessile and stemclasping and often auriculate at the base; *inflorescence* a ± lax or compact corymbose panicle; *capitula* on short peduncles, the involucres cylindrical, green or often purplish; *phyllaries* 9–13, in one row with a few small bracts at the base, 6–6.5 mm long, narrowly triangular, 2-nerved with narrow scarious margins, the acuminate tips ciliolate and usually slightly spreading; *outer florets* pistillate, the corollas filiform, 4-toothed; *disc florets* hermaphrodite, the corollas dilated above, 5-toothed; *achenes* 2.5–3 mm long, fusiform-cylindrical, slightly arcuate, narrowed to a very short beak above, strongly ribbed, usually with short white hairs in the grooves; *pappus bristles* soft and silky, 5–6 mm long.

Distribution   Alpine and subalpine tracts of the Australian Alps and associated ranges as far north as the A.C.T., and mountains of Tasmania.

Notes etc.   Occasional in tall alpine herbfields.

Reference: Belcher (1956).

S. pectinatus DC., Prodr. 6: 372 (1838)   'Alpine groundsel'

Stoloniferous perennial herb ± 10–20(–30) cm high; *flowering stem* glabrous or [348]
with short weak ± appressed septate hairs; *leaves* mostly radical, thick and slightly
fleshy, oblanceolate to oblong-spathulate in outline, pinnatifid to pinnately or
crenately lobed, narrowed to a long or short pseudo-petiole, stem-clasping and
usually reddish purple at the base, the lower cauline leaves similar but decreasing
in size up the stem and grading into acute or acuminate linear bracts; *capitula*
large, solitary, yellow, ± 2–4 cm diam.; *phyllaries* in several rows, the inner 8–10
mm long, herbaceous with broad scarious margins, papillose at the tips, the outer
phyllaries smaller and grading into the upper stem bracts; *ligules* of the ray florets
1–1.5 cm long, narrowly obovate or elliptic; *achenes* glabrous, 2.5–3.5 mm long;
pappus minutely barbellate, 5–7 mm long.

Alpine and subalpine tracts of the Australian Alps and mountains of Tasmania.   *Distribution*

Common in tall alpine herbfields and sod tussock grasslands; a small, apparently   *Notes and*
distinct ecotype occurs in *Epacris–Chionohebe* feldmark.   *Habitat*

S. lautus Forst. f. ex Willd. subsp. *alpinus* Ali, Aust. J. Bot. 17: 167 (1969)   'Variable groundsel'

Perennial herb about 20–45 cm high; *stems* usually much-branched, with sparse [349]
weak septate hairs or almost glabrous, erect or ascending, shortly creeping and
rooting at the base; *leaves* cauline, variable, the lower leaves petiolate, spathulate,
3–6(–8) cm long, coarsely serrate-dentate, the upper leaves becoming more dissected and ± elliptic in outline, pinnatifid or pinnatisect with entire or toothed
lobes, the uppermost leaves sessile with the small auriculate basal lobes ± stem-
clasping; *capitula* radiate, ± 1.5–2.5 cm diam., usually numerous in a loose subcorymbose inflorescence, rarely solitary in depauperate plants; *involucre*
campanulate, 4–6 cm long; *phyllaries* in one row with small linear fimbriate bracts
around the base, oblong-lanceolate to narrowly triangular, acute or acuminate
with brown ciliate tips, broad green centres and narrow overlapping scarious
margins; *ray florets* 10–15, pistillate, the ligules yellow, 5–8 mm long, elliptic
or obovate, blunt; *disc florets* hermaphrodite; *achenes* slender, 2–3 mm long,
slightly arcuate, grooved, glabrous or with lines of short antrorsely appressed
white hairs along the grooves; *pappus bristles* silky, minutely barbellate, ± 5 mm
long, deciduous; $n = 20^*$.

Alpine and subalpine tracts of the Australian Alps and associated ranges as far   *Distribution*
north as the A.C.T., and mountains of Tasmania.

Common in tall alpine herbfields and sod tussock grassland, especially on semi-   *Notes and*
bare areas between tussocks. Like many other alpine herbs, *S. lautus* has become   *Habitat*
more common since the cessation of grazing. The masses of yellow inflorescences
present spectacular displays in January. Ali (1969) notes that the forms with
glabrous and pubescent achenes occur in about equal proportions throughout
the range of this subspecies, and both forms are to be found above the treeline.
Ornduff (1964) suggests that the New Zealand forms of this species complex
may have been derived from populations originating in Australia.

References: Ali (1969); Ornduff (1960, 1964*).

Microseris D. Don

M. lanceolata (Walp.) Sch.-Bip., Pollichia 22–24: 310 (1866)
'Native dandelion'

[350] Perennial scapigerous herb about 10–40 cm high with a short vertical, usually sparingly branched rootstock and 1–10 fleshy cylindrical roots ± 4 mm diam., the leaves and scapes appearing glabrous but usually with scattered mealy hairs especially when young; *hairs* microscopic, septate, with an expanded terminal cell which is globular at first, becoming cupuliform or spathulate; *leaves* radical, ± 10–20(–30) cm long, very variable, from linear-lanceolate to oblanceolate or obovate-spathulate in outline, entire, distally toothed, or pinnatifid with narrow, often deflexed lobes; *scapes* long, hollow, naked or occasionally with a few small linear bracts; *capitula* solitary, ± 3–5 cm diam., nodding in bud; *involucre* cylindrical, 1.5–2.5 cm long; *phyllaries* green or with purplish midribs, in two series: outer series usually calyx-like, unequal, the longest about half as long as the inner series, ovate-triangular; inner series subequal, usually narrowly elliptic or ovate with pale membranous margins, ± glabrous or usually pubescent with dark hairs especially towards the tips; *florets* all ligulate, yellow, truncate and 5-toothed at the apex, the outer up to twice as long as the involucre; *achenes* cylindrical, 6–10 mm long, 10-ribbed, glabrous or sometimes sparsely covered with short non-septate pubescence; *pappus* 10–15 mm long, of (10–)15–20 thin scales which taper above to a long barbellate awn.

Distribution Temperate Australia, the alpine form of the species occurring in the Southern Tablelands of New South Wales and the Eastern Highlands of Victoria.

Notes and Habitat The alpine form of this variable species is often more robust, with larger and brighter yellow capitula than the forms at lower altitudes. *Microseris* tends to flower later than most other alpine species, and in most years the yellow capitula, carried on long stalks, are conspicuous in the tall alpine herbfields and sod tussock grassland during February and March. The genus has an interesting distribution in western North America, western South America, temperate Australia and New Zealand, and Raven (1972) maintains that this distribution must have been due to long-distance dispersal. The New Zealand species, *M. scapigera* (Sol. ex A. Cunn.) Sch.-Bip., is closely allied to *M. lanceolata*. The infraspecific taxonomy of *M. lanceolata* is currently being investigated by Sneddon (1977), who kindly provided the above description of the alpine form.

References: Chambers (1955); Raven (1972); Sneddon (1977).

Bibliography

Ali, S. I. (1969). *Senecio lautus* complex in Australia. V. Taxonomic interpretations. *Aust. J. Bot.* **17,** 161–76.

Allan, H. H. (1961). 'Flora of New Zealand.' Vol. 1. 1085 pp. (Govt. Printer: Wellington, N.Z.)

Andrews, A. E. J. (1973). Further light on the summit: Mt William IV not Mt Kosciusko. *J. R. Aust. Hist. Soc.* **59,** 114–27.

Aston, H. I. (1973). 'Aquatic Plants of Australia.' 368 pp. (Melbourne University Press.)

Barrow, M. D., Costin, A. B., and Lake, P. (1968). Cyclical changes in an Australian fjaeldmark community. *J. Ecol.* **56,** 89–96.

Beadle, N. C. W. (1971). 'Student's Flora of North Eastern New South Wales. Pt 1, Pteridophytes.' pp. 1–69. (University of New England Printery: Armidale, N.S.W.)

Belcher, R. O. (1956). A revision of the genus *Erechtites* (Compositae) with enquiries into *Senecio* and *Arrhenechthites*. *Ann. Mo. Bot. Gard.* **43,** 1–85.

Bergersen, F. J., and Costin, A. B. (1964). Root nodules on *Podocarpus lawrencei* and their ecological significance. *Aust. J. Biol. Sci.* **17,** 44–8.

Beuzenberg, E. J. (1961). Observations on sex differentiation and cytotaxonomy of the New Zealand species of the Hymenantherineae (Violaceae). *N.Z. J. Sci.* **4,** 337–49.

Billings, W. D. (1973). Arctic and alpine vegetations: similarities, differences, and susceptibility to disturbance. *BioScience* **23,** 697–704.

Blake, S. T. (1940). Notes on Australian Cyperaceae. III. *Proc. R. Soc. Queensl.* **51,** 32–50.

Blake, S. T. (1941). Notes on Australian Cyperaceae. V. *Proc. R. Soc. Queensl.* **52,** 55–61.

Blake, S. T. (1947). Notes on Australian Cyperaceae. VII. *Proc. R. Soc. Queensl.* **58,** 35–50.

Blake, S. T. (1969). Studies in Cyperaceae. *Contrib. Queensl. Herb.* No. 8, 1–48.

Blake, S. T. (1972). *Plinthanthesis* and *Danthonia* and a review of the Australian species of *Leptochloa* (Gramineae). *Contrib. Queensl. Herb.* No. 14, 1–19.

Bliss, L. C. (1971). Arctic and alpine plant life cycles. *Annu. Rev. Ecol. Syst.* **2,** 405–38.

Briggs, B. G. (1959). *Ranunculus lappaceus* and allied species of the Australian mainland. I. Taxonomy. *Proc. Linn. Soc. N.S.W.* **84,** 295–324.

Briggs, B. G. (1962). Interspecific hybridisation in the *Ranunculus lappaceus* group. *Evolution* **16,** 372–90.

Briggs, B. G. (1966). Chromosome numbers of some Australian monocotyledons. *Contrib. N.S.W. Natl Herb.* **4,** 24–34.

Briggs, B. G. (1973). Chromosomal studies in *Plantago* in Australia. *Contrib. N.S.W. Natl Herb.* **4**, 399–405.

Briggs, B. G., Carolin, R. C., and Pulley, J. M. (1973). New species and lecto-typification in Australian *Plantago*. *Contrib. N.S.W. Natl Herb.* **4**, 395–8.

Briggs, B. G., Carolin, R. C., and Pulley, J. M. (1977). Flora of New South Wales No. 181, Plantaginaceae, 1–35.

Briggs, B. G., and Ehrendorfer, F. (1976). *Chionohebe* a new name for *Pygmea* Hook. f. (Scrophulariaceae). *Contrib. Herb. Aust.* No. 25, 1–4.

Brock, R. D., and Brown, J. A. M. (1961). Cytotaxonomy of Australian *Danthonia*. *Aust. J. Bot.* **9**, 62–91.

Browne, W. R. (1965). The geology of Kosciusko. *Aust. Nat. Hist.* **15**, 56–60.

Browne, W. R. (1967). Geomorphology of the Kosciusko block and its north and south extensions. *Proc. Linn. Soc. N.S.W.* **92**, 117–44.

Bryant, W. G. (1971). The problem of plant introduction for alpine and sub-alpine regeneration, Snowy Mountains, N.S.W. *J. Soil Conserv. Serv. N.S.W.* **27**, 209–26.

Burbidge, N. T. (1958). A monographic study of *Helichrysum* subgenus *Ozothamnus* (Compositae) and of two related genera formerly included therein. *Aust. J. Bot.* **6**, 229–84.

Burbidge, N. T., and Gray, M. (1970). 'Flora of the Australian Capital Territory.' 447 pp. (Australian National University Press: Canberra.)

Burrows, C. J. (1960). Studies in *Pimelea*. I. The breeding system. *Trans. R. Soc. N.Z.* **88**, 29–45.

Byles, B. U. (1932). A reconnaissance of the mountainous part of the River Murray catchment in New South Wales. Commonw. Forestry Bur. Bull. No. 13, 1–34.

Cabrera, A. L. (1966). The genus *Lagenophora* (Compositae). *Blumea* **14**, 285–308.

Caine, N., and Jennings, J. N. (1968). Some blockstreams of the Toolong Range, Kosciusko State Park, New South Wales. *J. Proc. R. Soc. N.S.W.* **101**, 93–103.

Carolin, R. C. (1965). The genus *Geranium* in the south western Pacific area. *Proc. Linn. Soc. N.S.W.* **89**, 326–61.

Carolin, R. C. (1967). Geraniaceae. *Contrib. N.S.W. Natl Herb. Flora Ser.* No. 102, 1–23.

Cerceau-Larrival, M.-T. (1974). Palynologie et répartition des Ombellifères Australes actuelles. Relations avec les géoflores Tertiaires. *Sci. Geol. Bull. Ins. Geol. Uni. Louis Pasteur Strasbourg* **27**, 117–34.

Chambers, K. L. (1955). A biosystematic study of the annual species of *Microseris*. *Contrib. Dudley Herb.* **4**, 207–312.

Chapman, A. D. (1976). Name changes in *Kunzea* (Myrtaceae). *Contrib. Herb. Aust.* No. 18, 1–2.

Common, I. F. B. (1954). A study of the ecology of the adult Bogong moth *Agrotis infusa* (Boisd.) (Lepidoptera: Noctuidae), with special references to its behaviour during migration and aestivation. *Aust. J. Zool.* **2**, 223–62.

Conert, H. J. (1975). Die *Chionochloa* – arten von Australien und Neuguinea (Poaceae: Arundinoideae). *Senckenbergiana Biol.* **56**, 153–64.

Costin, A. B. (1954). 'A Study of the Ecosystems of the Monaro Region of New South Wales.' 860 pp. (Govt. Printer: Sydney.)

Costin A. B. (1958). The grazing factor and the maintenance of catchment values in the Australian Alps. CSIRO Aust. Div. Plant Ind. Tech. Pap. No. 10, 1–14.

Costin, A. B. (1961). Ecology of the high plains. I. *Proc. R. Soc. Vict.* **75**, 327–37.

Costin A. B. (1968). Alpine ecosystems of the Australasian Region. In 'Arctic and Alpine Environments', ed. W. H. Osburn and H. E. Wright, pp. 55–87. (Indiana University Press: Bloomington.)

Costin, A. B. (1972). Carbon-14 dates from the Snowy Mountains area, south-eastern Australia, and their interpretation. *Quat. Res. (N.Y.)* **2,** 579–90.

Costin, A. B., and Polach, H. A. (1971). Slope deposits in the Snowy Mountains, south-eastern Australia. *Quat. Res. (N.Y.)* **1,** 228–35.

Costin, A. B., Thom, B. G., Wimbush, D. J., and Stuiver, M. (1967). Nonsorted steps at Mount Kosciusko, Australia. *Bull. Geol. Soc. Am.* **78,** 979–92.

Costin, A. B., and Wimbush, D. J. (1973). Frost cracks and earth hummocks at Kosciusko, Snowy Mountains, Australia. *Arct. Alp. Res.* **5,** 111–20.

Costin, A. B., Wimbush, D. J., Barrow, M. D., and Lake, P. (1969). Development of soil and vegetation climaxes in the Mount Kosciusko area, Australia. *Vegetatio* **18,** 273–88.

Costin, A. B., Wimbush, D. J., Kerr, D., and Gay, L. W. (1959). Studies in catchment hydrology in the Australian Alps. I. Trends in soils and vegetation. CSIRO Aust. Div. Plant Ind. Tech. Pap. No. 13, 1–36.

Curtis, W. M. (1956). 'The Student's Flora of Tasmania. Pt 1. (Gymnospermae and Angiospermae: Ranunculaceae to Myrtaceae).' 234 pp. (Govt. Printer: Hobart.)

Curtis, W. M. (1963), 'The Student's Flora of Tasmania. Pt 2. (Lythraceae to Epacridaceae).' 475 pp. (Govt. Printer: Hobart.)

Curtis, W. M. (1967). 'The Student's Flora of Tasmania. Pt 3. (Plumbaginaceae to Salicaceae).' 661 pp. (Govt Printer: Hobart.)

Daubenmire, R. (1954). Alpine timberlines in the Americas and their interpretation. *Butler Univ. Bot. Stud.* **11,** 119–36.

Davis, G. L. (1948). Revision of the genus *Brachycome* Cass. I. Australian species. *Proc. Linn. Soc. N.S.W.* **73,** 142–241.

Davis, G. L. (1949). Revision of the genus *Brachycome* Cass. Pt III. Description of three new Australian species and some new locality records. *Proc. Linn. Soc. N.S.W.* **74,** 145–52.

Davis, G. L. (1950). A revision of the Australian species of the genus *Lagenophora* Cass. *Proc. Linn. Soc. N.S.W.* **75,** 122–32.

Davis, G. L. (1956). Revision of the genus *Podolepis* Labill. *Proc. Linn. Soc. N.S.W.* **81,** 245–86.

Dawson, J. W. (1961). A revision of the genus *Anisotome* (Umbelliferae). *Univ. Calif. Berkeley Publ. Bot.* **33,** 1–98.

Dawson, J. W. (1967). The New Zealand species of *Gingidium* (Umbelliferae). *N.Z. J. Bot.* **5,** 84–116.

Dawson, J. W. (1971). Relationships of the New Zealand Umbelliferae. In 'The Biology and Chemistry of the Umbelliferae' (*Bot. J. Linn. Soc.* **64,** suppl. I), ed. V. H. Heywood, pp. 43–61. (Academic Press: London.)

Dawson, J. W. (1974). Validation of *Gingidia* (Umbelliferae). *Kew Bull.* **29,** 476.

Dawson, J. W. (1976). The Australian species of *Seseli* L. transferred to *Gingidia* Dawson. *Contrib. Herb. Aust.* No. 23, 1–2.

Denton, G. H., Armstrong, R. L., and Stuiver, M. (1971). The Late Cenozoic glacial history of Antarctica. In 'Late Cenozoic Glacial Ages', ed. K. K. Turekian, pp. 267–306. (Yale University Press: New Haven.)

Drury, D. G. (1972). The cluster and solitary-headed cudweeds native to New

Zealand: (*Gnaphalium* section Euchiton–Compositae). *N.Z. J. Bot.* **10,** 112–79.

Drury, D. G. (1974). A broadly based taxonomy of *Lagenifera* section *Lagenifera* and *Solenogyne* (Compositae–Astereae), with an account of their species in New Zealand. *N.Z. J. Bot.* **12,** 365–96.

Edgar, E. (1964). Leaf characters and a new species in *Oreobolus* (Cyperaceae). *N.Z. J. Bot.* **2,** 454–8.

Edgar, E. (1975). Australasian *Luzula*. *N.Z. J. Bot.* **13,** 781–802.

Erickson, R. (1958). 'Triggerplants.' 229 pp. (Paterson Brokensha: Perth.)

Erickson, R. (1968). 'Plants of Prey in Australia.' 94 pp. (Lamb Publs Ltd, W.A.)

Falvey, D. A. (1972). Plate tectonics: A dynamic approach to modern geological theory. *Aust. Nat. Hist.* **17,** 258–64.

Fisher, F. J. F. (1965). The Alpine Ranunculi of New Zealand. N.Z. Dep. Sci. Ind. Res. Bull. No. 165, 1–192.

Flood, J. M. (1973). The moth hunters. Investigations towards a prehistory of the south-eastern highlands of Australia. Ph. D. thesis, Australian National University, Canberra.

Galloway, R. W. (1963). Glaciation in the Snowy Mountains: A reappraisal. *Proc. Linn. Soc. N.S.W.* **88,** 180–98.

Gandoger, M. (1899). Note sur la flore du Mont Kosciusko (Australie Meridionale). *Bull. Soc. Bot. Fr.* **46,** 391–4.

Given , D. R. (1969). Taxonomic notes on the genus *Celmisia* (Compositae). *N.Z. J. Bot.* **7,** 389–99.

Gray, M. (1974). Miscellaneous notes on Australian plants: *Erigeron, Uncinia, Poa* and *Erythranthera. Contrib. Herb. Aust.* No. 6, 1–5.

Gray, M. (1976). Miscellaneous notes on Australian plants. 3. *Craspedia, Gnaphalium, Epacris, Tasmannia, Colobanthus* and *Deyeuxia. Contrib. Herb. Aust.* No. 26, 1–11.

Gray, N. E. (1956). A taxonomic revision of *Podocarpus* X, The South Pacific species of section Eupodocarpus, subsection D. *J. Arnold Arbor. Harv. Univ.* **37,** 160–72.

Green, P. S. (1970). Notes relating to the Floras of Norfolk and Lord Howe Islands. I. *J. Arnold Arbor. Harv. Univ.* **51,** 204–20.

Hamlin, B. G. (1959). A revision of the genus *Uncinia* (Cyperaceae–Caricoideae) in New Zealand. *Bull. Dom. Mus. (Wellington)* **19,** 1–106.

Hancock, W. K. (1972). 'Discovering Monaro. A Study of Man's Impact on his Environment.' 209 pp. (Cambridge University Press.)

Healy, A. J. (1948). Contributions to a knowledge of the naturalized flora of New Zealand. 2. *Trans. R. Soc. N.Z.* **77,** 172–85.

Helms, R. (1893). Report on the grazing leases of the Mount Kosciusko plateau. *Agric. Gaz. N.S.W.* **4,** 530–1.

Heywood, V. H. (1971). Flora Europaea – a progress report, 1967–1970. *Boissiera* **19,** 17–20.

Hill, A. W. (1918). The genus *Caltha* in the southern hemisphere. *Ann. Bot. (Lond.)* **32,** 421–35.

Hooker, J. D. (1860). 'Introductory Essay to Flora Tasmaniae.' pp. i–cxxviii. (Lovell Reeve: London.)

Hotchkiss, A. T. (1955). Chromosome numbers and pollen tetrad size in Winteraceae. *Proc. Linn. Soc. N.S.W.* **80,** 47–53.

Hulten, E. (1959). The *Trisetum spicatum* complex. *Sven. Bot. Tidskr.* **53**, 203–28.

Ives, J. D., and Barry, R. G. (eds) (1974). 'Arctic and Alpine Environments.' 999 pp. (Methuen: London.)

Jacobs, M., and Moore, D. M. (1971). *Flora Malesiana Ser. 1* (Violaceae), **7**, 179–212.

Jeans, D. N., and Gilfillan, W. G. R. (1969). Light on the summit: Mount William the Fourth or Kosciusko? *J. R. Aust. Hist. Soc.* **55**, 1–18.

Johnson, L. A. S., and Evans, O. D. (1966). Restionaceae. *Contrib. N.S.W. Natl Herb. Flora Ser.* No. 25, 2–28.

Jones, D. L. (1972). The pollination of *Prasophyllum alpinum* R. Br. *Victorian Nat.* **89**, 260–3.

Jones, J. G. (1971). Australia's Caenozoic drift. *Nature (Lond.)* **230**, 237–9.

Kern, J. H. (1974). *Flora Malesiana Ser. 1* (Cyperaceae), **7**(3), 435–753.

Lourteig, A. (1968). Révision de *Juncus* subgen. Septati Buch. *CNFRA (Biol.)* **23**, 33–49.

Mackerras, I. M. (1970). Composition and distribution of the fauna. In 'The Insects of Australia', pp. 187–203 (Melbourne University Press.)

McLuckie, J., and Petrie, A. H. K. (1927). The vegetation of the Kosciusko Plateau. Part 1. The plant communities. *Proc. Linn. Soc. N.S.W.* **52**, 187–221.

Maiden, J. H. (1894). A list of plants collected by Mr Richard Helms in the Australian Alps, February 1893. *Agric. Gaz. N.S.W.* **5**, 836–41.

Maiden, J. H. (1898). A contribution towards a flora of Mount Kosciusko. *Agric. Gaz. N.S.W.* **9**, 720–40.

Maiden, J. H. (1899). A second contribution towards a flora of Mount Kosciusko. *Agric. Gaz. N.S.W.* **10**, 1001–42.

Maiden, J. H. (1904). The tree line in the Australian Alps. *Victorian Nat.* **20**, 84.

Mark, A. F., and Adams, Nancy M. (1973). 'New Zealand Alpine Plants.' 262 pp. (A. H. and A. W. Reed: Wellington, N.Z.)

Mathias, M. E., and Constance, L. (1955). The genus *Oreomyrrhis* (Umbelliferae). *Univ. Calif. Berkeley Publ. Bot.* **27**, 347–416.

Mathias, M. E., and Constance, L. (1971). A first revision of *Huanaca* (Umbelliferae–Hydrocotyloideae). *Kurtziana* **6**, 7–23.

Melville, R. (1955). Some Ranunculi of Tasmania and south-eastern Australia. *Kew Bull.* 1955, 193–220.

Moar, N. T. (1966). Studies in pollen morphology. 3. The genus *Gingidium* J. R. et G. Forst. in New Zealand. *N.Z. J. Bot.* **4**, 322–32.

Moore, D. M. (1963). The status of *Viola betonicifolia* Sm. in New Guinea. *Feddes Repert.* Bd. 68, H.2, 81–6.

Moore, D. M. (1964). Experimental taxonomic studies in Antarctic Floras. In 'Biologie Antarctique' (Proc. 1st Symp. Antarctic Biol., Paris, 1962), ed. R. Carrick *et al.*, pp. 195–202. (Hermann: Paris.)

Moore, D. M. (1970). Studies in *Colobanthus quitensis* (Kunth.) Bartl. and *Deschampsia antarctica* Desv. II. Taxonomy, distribution and relationships. *Br. Antarct. Surv. Bull.* **23**, 63–80.

Moore, D. M., and Chater, A. O. (1971). Studies in bipolar disjunct species. 1. *Carex. Bot. Not.* **124**, 317–34.

Moore, L. B. (1966). Australasian Asteliads (Liliaceae); with special reference to New Zealand species of *Astelia* subgen. Tricella. *N.Z. J. Bot.* **4**, 201–40.

Moore, L. B., and Edgar, E. (1970). 'Flora of New Zealand.' Vol. II, 354 pp. (Govt. Printer: Wellington, N.Z.)

Morris, P. F. (1924). A new species of *Brachycome. Victorian Nat.* **41**, 31.

Mueller, F. (1855*a*). Botany of Victoria (Southern Australia). Extracts of letters from Dr Mueller, Colonial Botanist, Victoria. *Hook. Kew J.* **7**, 233–42.

Mueller, F. (1855*b*). Descriptive characters of new alpine plants from Continental Australia. *Trans. Phil. Soc. Vict.* **1**, 96–111.

Mulvaney, D. J., and Golson, J. (eds) (1971). 'Aboriginal Man and Environment in Australia.' 389 pp. (Australian National University Press: Canberra.)

National Parks and Wildlife Service of New South Wales (1974). 'Kosciusko National Park Plan of Management.' 55 pp. (Nat. Parks and Wildl. Serv.: Sydney.)

Nelmes, E. (1944). A key to the Australian species of *Carex* (Cyperaceae). *Proc. Linn. Soc. Lond. (sess. 155, 1942–3)*, 277–85.

Nicholls, W. H. (1969). 'Orchids of Australia', ed. D. L. Jones and T. B. Muir. 141 pp. (Nelson: Melbourne.)

Nilsson, Ö. (1966*a*). Studies in *Montia* L. and *Claytonia* L. and allied genera. I. Two new genera *Mona* and *Paxia. Bot. Not.* **119**, 265–85. (Also correction to *Neopaxia* l.c. p. 469.)

Nilsson, Ö. (1966*b*). Studies in *Montia* L. and *Claytonia* L. and allied genera. II. Some chromosome numbers. *Bot. Not.* **119**, 464–8.

Nilsson, Ö. (1967). Studies in *Montia* L. and *Claytonia* L. and allied genera. III. Pollen morphology. *Grana Palynol.* **7**, 279–363.

Nordenskiöld, H. (1969). The genus *Luzula* in Australia. *Bot. Not.* **122**, 69–89.

Nordenskiöld, H. (1971). Hybridization experiments in the genus *Luzula*. IV. Studies with taxa of the *campestris–multiflora* complex from the Northern and Southern Hemispheres. *Hereditas* **68**, 47–60.   [1–207.

Oliver, W. R. B. (1935). The genus *Coprosma. Bull. Bernice P. Bishop Mus.* **132**,

Oliver, W. R. B. (1956). The genus *Aciphylla. Trans. R. Soc. N.Z.* **84**, 1–18.

Orchard, A. E. (1975). Taxonomic revisions in the family Haloragaceae. 1. The genera *Haloragis, Haloragodendron, Glischrocaryon, Meziella* and *Gonocarpus. Bull. Auckland Inst. Mus.* **10**, 1–299.

Ornduff, R. (1960). An interpretation of the *Senecio lautus* complex in New Zealand. *Trans. R. Soc. N.Z.* **88**, 63–77.

Ornduff, R. (1964). Evolutionary pathways of the *Senecio lautus* alliance in New Zealand and Australia. *Evolution* **18**, 349–60.

Osburn, W. H., and Wright, H. E. (eds) (1968). 'Arctic and Alpine Environments.' 308 pp. (Indiana University Press: Bloomington.)

Parris, B. S. (1975). A revision of the genus *Grammitis* Sw. (Filicales: Grammitidaceae) in Australia. *Bot. J. Linn. Soc.* **70**, 21–43.

Petrie, D. (1890). Descriptions of new native plants. *Trans. Proc. N.Z. Inst.* **22**, 439–43.

Philipson, W. R. (1972). The generic status of the southern hemisphere gentians. *Adv. Pl. Morph. Meerut, India. Sarita Prakashan*, 417–22.

Ratkowsky, D., and Ratkowsky, A. (1974). New plant discoveries in Tasmania. *Aust. Plants* **7**, 384–6.

Raunkiaer, C. (1934). 'The Life-forms of Plants and Statistical Plant Geography.' 632 pp. (Clarendon Press: Oxford.)

Raven, P. H. (1963). The generic position of *Boisduvalia tasmanica. Aliso* **5**, 247–9.

Raven, P. H. (1972). Plant species disjunctions: a summary. *Ann. Mo. Bot. Gard.* **59**, 234–46.

Raven, P. H. (1973). Evolution of subalpine and alpine plant groups in New Zealand. *N.Z. J. Bot.* **11,** 177–200.

Raven, P. H., and Raven, T. E. (1976). The genus *Epilobium* (Onagraceae) in Australasia: a systematic and evolutionary study. N.Z. DSIR Bull. No. 216, 1–321. [307–15.

Rothmaler, W. (1955). *Alchemilla* und *Aphanes* in Australien. *Feddes Repert.* **58,**

Rupp, H. M. R. (1969). 'The Orchids of New South Wales.' Facsimile 1969, ed., with suppl., D. J. McGillivray. 177 pp. (Govt Printer: Sydney.)

Salmon, J. T. (1968). 'Field Guide to the Alpine Plants of New Zealand.' 327 pp. (Reed: Wellington, N.Z.)

Sealy, J. R. (1950). *Prostanthera cuneata. Curtis Bot. Mag.* **167,** t. 132.

Shaw, H. K. A., and Turrill, W. B. (1928). Asperulae Australienses. *Kew Bull. 1928,* 81–105.

Skottsberg, C. (1934). Studies in the genus *Astelia* Banks et Solander. *K. Sven. Vetenskapsakad. Handl. Ser. 3,* **14**(2), 1–106.

Smit, P. G. (1973). A revision of *Caltha* (Ranunculaceae). *Blumea* **21,** 119–50.

Smith-White, S. (1955). Chromosome numbers and pollen types in the Epacridaceae. *Aust. J. Bot.* **3,** 48–67.

Smith-White, S. (1959). Pollen development patterns in the Epacridaceae. A problem in cytoplasmic-nucleus interaction. Presidential address. *Proc. Linn. Soc. N.S.W.* **84,** 8–35.

Smith-White, S., Carter, C. R., and Stace, H. M. (1970). The cytology of *Brachycome.* I. The subgenus *Eubrachycome* – a general survey. *Aust. J. Bot.* **18,** 99–125.

Sneddon, B. V. (1977). A biosystematic study on *Microseris* subgen. *Monermos.* Ph.D. thesis, Victoria University of Wellington, New Zealand.

Stace, H. M. (1974). Cytogenetic studies in *Calotis* and *Brachycome*. Ph.D. thesis, University of Sydney.

Stauffer, H. U. (1959). Revisio Anthobolearum. *Mitt. Bot. Mus. Univ. Zürich* **213,** 1–260.

Stirling, J. (1885). Remarks on the flora of the Australian Alps, with introductory notes on the geology and meteorology of the area. *Wing's Southern Science Record* **1** (No. 1) 10–12, (No. 2) 30–3, (No. 3) 50–4, (No. 4) 76–80, (No. 5) 92–4. (Nos 4 and 5 are relevant to the Kosciusko flora.)

Stirling, J. (1887). Notes on the flora of Mount Hotham. *Victorian Nat.* **4,** 72–8.

Thompson, J. (1961). Papilionaceae (in part). *Contrib. N.S.W. Natl Herb., Flora Ser.* No. 101(1), 1–91.

Thurling, N. (1966). Population differentiation in Australian *Cardamine*. I. Variation in leaf characters. *Aust. J. Bot.* **14,** 167–78.

Thurling, N. (1966b). Population differentiation in Australian *Cardamine*. II. Response to variations in temperature and light intensity. *Aust. J. Bot.* **14,** 179–88.

Thurling, N. (1966c). Population differentiation in Australian *Cardamine*. III. Variation in germination response. *Aust. J. Bot.* **14,** 189–94.

Thurling, N. (1968). A cytotaxonomic study of Australian *Cardamine*. *Aust. J. Bot.* **16,** 515–23.

Tindale, M. D. (1955a). Studies in Australian Pteridophytes. 2. *Victorian Nat.* **71,** 191–2.

Tindale, M. D. (1955b). Some notes on the genus *Polystichum* in south-eastern Australia. *Proc. Linn. Soc. N.S.W.* **80,** 54–6.

Tindale, M. D. (1961). Pteridophyta of south-eastern Australia. *Contrib. N.S.W. Natl Herb., Flora Ser.* Nos 208–211, 1–78.

Totterdell, C. J., and Nebauer, N. R. (1973). Colour aerial photography in the reappraisal of alpine soil erosion. *J. Soil Conserv. Serv. N.S.W.* **29**, 130–58.

Vickery, J. W. (1940). A revision of the Australian species of *Deyeuxia* Clar. ex Beauv. with notes on the status of the genera *Calamagrostis* and *Deyeuxia. Contrib. N.S.W. Natl Herb.* **1**(2), 43–82.

Vickery, J. W. (1941). A revision of the Australian species of *Agrostis* Linn. *Contrib. N.S.W. Natl Herb.* **1**(3), 101–19.

Vickery, J. W. (1956). A revision of the Australian species of *Danthonia* DC. *Contrib. N.S.W. Natl Herb.* **2**(3), 249–325.

Vickery, J. W. (1970). A taxonomic study of the genus *Poa* L. in Australia. *Contrib. N.S.W. Natl Herb.* **4**(4), 145–243.

Vickery, J. W. (1975). Flora of New South Wales No. 19, Gramineae (supplement to Part 1, Part 2), 125–306.

Vink, W. (1970). The Winteraceae of the Old World. I. *Pseudowintera* and *Drimys* —morphology and taxonomy. *Blumea* **18**, 225–354.

Wakefield, N. A. (1955). 'Ferns of Victoria and Tasmania.' 71 pp. (Field Nat. Club Victoria.)

Walker, P. H., and Costin, A. B. (1971). Atmospheric dust accession in south-eastern Australia. *Aust. J. Soil Res.* **9**, 1–5.

Weindorfer, G. (1904a). Some comparison of the alpine flora of Australia and Europe. *Victorian Nat.* **20**, 64–70.

Weindorfer, G. (1904b). Some considerations of the origin of our alpine flora. *Victorian Nat.* **21**, 6–9.

Willis, J. H. (1954). Two new Victorian species of alpine *Podolepis. Victorian Nat.* **70**, 223–6.

Willis, J. H. (1956a). Additions to the Victorian sedge flora (Cyperaceae). *Victorian Nat.* **73**, 69–74.

Willis, J. H. (1956b). Two puzzling alpine heaths. *Victorian Nat.* **73**, 56–8.

Willis, J. H. (1957). Vascular flora of Victoria and South Australia. Sundry new species, varieties, combinations, records and synonymies. *Victorian Nat.* **73**, 188–202.

Willis, J. H. (1967). Systematic notes on the indigenous Australian flora. *Muelleria* **1**(3), 117–63.

Willis, J. H. (1970). 'A Handbook to Plants in Victoria. Vol. 1. Ferns, Conifers, and Monocotyledons.' 448 pp. (2nd Ed.) (Melbourne University Press.)

Willis, J. H. (1972, pub. 1973). 'A Handbook to Plants in Victoria. Vol. II. Dicotyledons.' 832 pp. (Melbourne University Press.)

Willis, M. (1949). 'By Their Fruits. A Life of Ferdinand von Mueller, Explorer and Botanist.' 187 pp. (Angus and Robertson: Sydney.)

Wilson, P. G. (1960). A consideration of the species previously included within *Helipterum albicans* (A. Cunn.) DC. *Trans. R. Soc. S. Aust.* **83**, 163–77.

Wilson, P. G. (1970). A taxonomic revision of the genera *Crowea, Eriostemon* and *Phebalium* (Rutaceae). *Nuytsia* **1**(1), 1–155.

Wimbush, D. J., and Costin, A. B. (1973). Vegetation mapping in relation to ecological interpretation and management in the Kosciusko alpine area. CSIRO Aust. Div. Plant Ind. Tech. Pap. No. 32, 1–22.

Yeo, P. F. (1968). The evolutionary significance of the speciation of *Euphrasia* in Europe. *Evolution* **22**, 736–47.

Yeo, P. F. (1973). Plants: Wild and Cultivated. B.S.B.I. Conf. Rep. No. 13, ed. P. S. Green. *Acaena,* pp. 51–5; The species of *Acaena* with spherical heads cultivated and naturalized in the British Isles, *lc.* Appendix III, pp. 193–221.

Zotov, V. D. (1963). Synopsis of the grass subfamily Arundinoideae in New Zealand. *N.Z. J. Bot.* **1,** 78–136.

Zotov, V. D. (1965). Grasses of the subantarctic islands of the New Zealand region. *Rec. Dom. Mus. Wellington* **5,** 101–46.

Zotov, V. D. (1973). *Hierochloe* R. Br. (Gramineae) in New Zealand. *N.Z. J. Bot.* **11,** 561–80.

Zwinger, A. H., and Willard, B. E. (1972). 'Land Above the Trees. A Guide to American Alpine Tundra.' 489 pp. (Harper and Row: New York.)

Glossary

abaxial – the side of a lateral organ away from the axis; dorsal.

acaulescent – stemless or apparently so.

achene – a small, dry, indehiscent 1-seeded fruit, as developed by species in the families Ranunculaceae, Compositae etc.

acicular – needle-shaped.

actinomorphic – radially symmetrical; with more than one plane of symmetry – regular; used in reference to flowers having petals and sepals of similar shape and size in each whorl (cf. zygomorphic).

acuminate – tapering to a prolonged point.

acute – distinctly and sharply pointed but not tapering.

adaxial – the side of a lateral organ towards the axis; ventral. (cf. abaxial).

adnate – united to an organ of a different kind, e.g. stamens with the corolla.

alternate – placed singly, at different heights along an axis; not opposite or whorled.

alveolus (pl. *alveoli*) – a small cavity; hence *alveolate* – with pits or depressions suggesting honeycomb.

amplexicaul – clasping or embracing the stem.

androgynous – with male and female flowers in the same inflorescence.

annual – completing its life-cycle in one year.

annulus – in ferns, a ring of thick-walled cells forming part of the opening mechanism of a sporangium.

anther – the part of the stamen containing the pollen grains.

anthesis – the act of flowering; the stage when pollen is shed.

antrorse – directed upward or forward. (cf. retrorse).

apiculate – with a small broad point at the apex.

appressed – lying flat against an organ but not united with it.

arcuate – moderately curved; bent like a bow.

articulate – jointed; provided with nodes or joints, or places where separation may naturally take place.

ascending – sloping or curving upwards.

attenuate – gradually tapering at base or apex.

auricle – an ear-shaped appendage or lobe, as at the base of a leaf; hence *auriculate* – with auricles.

awn – a slender bristle-like projection, often from the tip or back of an organ.

axil – upper angle between two dissimilar parts, e.g. between a petiole and stem; hence *axillary* – occurring in the axil.

barbellate – finely barbed; commonly used to describe the antrorsely scabrid pappus bristles of Compositae.

basifixed – attached or fixed by the base.

beak – a prominent pointed projection.

berry – a fleshy indehiscent fruit usually containing more than one seed, without a stony layer surrounding the seeds.

bifid – divided or cleft into two parts.

bisexual – having both sexes present and functional in the one flower.

biternate – compound ternate.

blunt – see obtuse.

bract – a reduced or modified leaf, e.g. the small scale-like leaves subtending a flower or belonging to an inflorescence; hence *bracteate* – with bracts; *bracteole* – a small bract.

caducous – falling off early or prematurely.

caespitose – growing in ± dense tufts.

callous – hardened and abnormally thickened.

callus – a thickened, usually hardened part; in grasses, the hard basal projection at the base of the floret or spikelet where these form the seed unit.

calyx – the sepals collectively, which comprise the outer whorl of the flower.

calyx tube – the tube of a gamosepalous calyx.

campanulate – bell-shaped.

canescent – becoming hoary, usually with a greyish pubescence.

capillary – hair-like, very slender.

capitate – in heads; aggregated into a dense or compact cluster.

capitulum (pl. *capitula*) – a dense head-like inflorescence of usually numerous, ± sessile flowers.

capsule – a dry dehiscent fruit composed of more than one carpel.

carpel – one unit of the gynoecium or female part of the flower, consisting of the ovule-bearing ovary, the style (when present) and the stigma.

carpophore – in Umbelliferae, a wiry stalk that supports each half (carpel) of the dehiscing fruit.

cartilaginous – tough and hard but not bony; gristly.

caruncle – a fleshy outgrowth or appendage at or near the hilum (point of attachment) of a seed.

caryopsis – an achene developed from a one-carpelled superior ovary in which the pericarp is united with the seed-coat, e.g. the grain or fruit of most grasses.

castaneous – chestnut-brown in colour.

caudex – the woody base of a perennial plant.

cauline – pertaining or belonging to the stem.

caulorhiza – rootstock; often used when the rootstock is relatively short and thick.

channelled – grooved longitudinally.

chartaceous – of papery texture.

cilia – a short fine hair; hence *ciliate* – said of a margin fringed with fine hairs resembling eyelashes; dimin. *ciliolate*.

circumsciss or circumscissile – opening or dehiscing along a horizontal line so that the top comes off like a lid.

clathrate – latticed.

clavate – club-shaped; said of a long body thickened towards one end.

claw – the narrowed stalk-like base of some sepals or petals.

cleistogamous – having fertilization occur within the unopened flower.

coherent – two or more similar parts or organs joined.

column – body formed by the union of stamens, style and stigmas, especially in Orchidaceae.

coma – a tuft of hairs at the end of some seeds, e.g. in *Epilobium*.

compound – of a leaf, made up of several distinct leaflets; of an inflorescence, with the axis branched.

compressed – flattened.

cone – a compact group of sporophylls borne on a central axis, as in pines and other gymnosperms.

confluent – merging or blending together.

congested – crowded together.

connate – referring to organs of the same kind growing together and becoming joined.

connective – the filament or tissue connecting the two cells of an anther.

connivent – coming together or converging but not organically connected.

contorted – of perianth lobes in bud, with each lobe overlapping the next with the same edge and appearing twisted.

contracted – said of inflorescences that are narrow and dense with short or appressed branches.

convolute – rolled up, the margins overlapping.

cordate – heart-shaped; ovate in outline with a notch at the base; sometimes only referring to the basal part rather than to the whole outline.

coriaceous – leathery texture.

corolla – referring to the petals as a whole.

corymb – a raceme with the pedicels becoming shorter towards the top, so that all the flowers are at approximately the same level; adj. *corymbose*.

crenate – having a margin with shallow rounded or blunt teeth; dimin. *crenulate* – finely crenate.

crispate – curled; an extreme form of undulate.

culm – the stem, particularly of grasses and sedges, bearing leaves and inflorescence.

cuneate – wedge-shaped, narrowest at point of attachment and increasing regularly in width to the apex.

cupuliform – cup-shaped.

cuspidate – tipped with a sharp rigid point.

cymbiform – boat-shaped.

cyme – an inflorescence in which the terminal flower terminates the growth of the main axis, which is then replaced by the growth of one or two lateral buds, the process being repeated throughout the development of the inflorescence.

deciduous – not persistent; falling off.

decumbent – reclining or lying on the ground but with the ends ascending.

decurrent – extending down and adnate to the stem, as in leaves where the blade is continued downwards as a wing on petiole or stem.

decussate – (of leaves) opposite with the successive pairs arranged at right angles to each other.

dehiscence – the method or process of opening of a seed pod or anther; hence *dehiscent* – opening to shed its seeds or spores.

deltoid – triangular.

dendritic – (of hairs) hairs that branch like a tree.

dentate – with sharp, spreading, rather coarse indentations or teeth that are perpendicular to the margin; dimin. *denticulate* – minutely or finely dentate.

depauperate – impoverished, starved, reduced in size or dwarfed.

dichotomous – branching by forking in one or more pairs.

didynamous – with 4 stamens in two pairs of two different lengths.

digitate – referring to a compound leaf whose leaflets spread from a common centre, like the fingers of a hand.

dioecious – unisexual, the male and female elements in different plants.

disc – the fleshy, sometimes nectar-secreting portion of the receptacle surrounding or surmounting the ovary; the common receptacle in the head of Compositae.

discolorous – of different colours, as on the upper and lower surfaces of a leaf.

distal – towards the free, as opposed to the attached or proximal, end of an organ.

distichous – conspicuously 2-ranked; in two opposite rows.

divaricate – spreading widely apart; extremely divergent.

dorsal – relating to the back; the surface turned away from the axis; abaxial.

dorsifixed – (of anthers) attached at or by the back.

drupe – a fleshy one-seeded indehiscent fruit with the seed enclosed in a stony endocarp; a stone fruit, e.g. a plum; hence *drupaceous* – resembling a drupe.

ebracteate – without bracts.

eglandular – without glands.

ellipsoid – a solid object elliptic in outline.

elliptic – in outline broadest across the middle and gradually narrowed to ± rounded ends.

emarginate – with a shallow notch at the apex.

endemic – native or confined naturally to a particular and usually restricted area or region.

entire – with a continuous margin, not toothed or divided.

epigynous – borne on the ovary; said of floral parts in which the ovary is inferior and not perigynous.

epipetalous – borne on or arising from the petals or corolla.

equitant – of distichous leaves folded longitudinally and overlapping in their lower parts, e.g. the leaves of *Iris*.

erecto-patent – spreading at an angle of about 45°.

erose – with irregular margin, as if bitten or gnawed.

excurrent – extending beyond the margin or tip, e.g. a midrib developing into a mucro or awn.

exserted – protruding.

exstipulate – without stipules.

extrorse – facing or opening outward.

falcate – sickle- or scythe-shaped.

fasciculate – in close bundles or clusters.

ferruginous – rust-coloured.

fertile – producing seed capable of germination; or (of anthers) containing viable pollen.

filament – the stalk of a stamen to which the anthers are attached; a thread-like structure.

filiform – thread-like, long and very slender.

fimbriate – fringed, the hairs longer or coarser as compared with ciliate; hence *fimbriolate* – very finely fimbriate.

flaccid – limp, not rigid.

flexuose – bent alternately in different directions; zig-zagged.

floret – a small flower, e.g. as in Compositae, or the lemma and palea with included flower in Gramineae.

floriferous – having an abundance of flowers.

folded – referring to leaf blades folded lengthwise about the midrib with the upper surface within.

foliaceous – leaflike.

follicle – a single carpellate dry fruit dehiscing along one line of suture.

fruit – the ripened ovary containing the seeds; often used to include other associated parts such as fleshy receptacle.

funnelform – with the tube gradually widening upward and passing insensibly into the limb.

fusiform – spindle-shaped; a solid, swollen in the middle and narrowed to both ends.

gamopetalous – with a corolla of one piece, the petals united at least at the base; sympetalous.

gamosepalous – with a calyx whose sepals are marginally connate, in whole or in part.

geniculate – bent, like a knee.

glabrescent – becoming glabrous with increasing age or maturity.

glabrous – without hairs of any sort.

gland – a secreting organ or any small vesicle containing oil, resin etc., sunk in, on the surface of or protruding from any part of a plant; hence *glandular* – having glands.

glaucous – bluish green, or covered with a waxy bloom which may rub off.

globular – shaped like a globe; spherical.

glume – a small chaffy or membranous bract, especially one of the two empty bracts at the base of most grass spikelets, or the bract subtending the flowers of sedges; hence *glumaceous* – resembling a glume.

grain – caryopsis or naked seed of grasses.

gymnosperm – a group of seed plants (e.g. pines) in which the ovules are not enclosed in an ovary.

gynodioecious – having female and hermaphrodite flowers on separate plants.

gynoecium – the female parts of the flower, made up of one or more ovaries with their styles and stigmas.

hastate – having the shape of an arrowhead but with the basal lobes spreading nearly or quite at right angles.

haustorium – a sucker, or food-absorbing organ of parasitic plants.

herb – any vascular plant that is not woody.

herbaceous – (of a plant organ) green, with the texture of leaves.

hermaphrodite – bisexual, i.e. with stamens and pistil in the same flower.

hirsute – covered with long, not interwoven hairs; this term has been used by various authors to denote either soft or stiff hairs; herein the term is qualified, e.g. softly hirsute, stiffly hirsute.

hispid – beset with rough hairs or bristles; dimin. *hispidulous.*

hoary – covered with a close white or whitish pubescence.

hyaline – thin, delicate and translucent or transparent.

hydathode – a water pore or water gland, often found at the tips of leaves.

hypanthium – the cuplike receptacle derived usually from the fusion of floral envelopes and androecium, and on which are seemingly borne calyx, corolla and stamens.

hypogynous – borne below the ovary.

imbricate – with edges overlapping, like tiles on a roof.

included – not protruding beyond the surrounding organ; not exserted.

incurved – bent gradually inwards.

indehiscent – not opening to release the seeds.

indumentum – a general term for a covering of hairs of any form.

indurated – hardened and toughened.

indusium (pl. *indusia*) – the tissue ± covering or enclosing the sori of some ferns; a cup enclosing the stigma in Goodeniaceae.

inferior (ovary) – the ovary apparently surrounded by and fused with the receptacle, the perianth being inserted around the top.

inflorescence – a general term for the collection of flowering parts of a flowering branch including its branches, bracts and flowers; the arrangement of the flowers or mode of flowering.

insectivorous – insect-catching, as in *Drosera.*

inserted – attached to or growing upon, e.g. as a stamen attached to the corolla.

internode – the part of a stem between two successive nodes or 'joints'.

involucre – a ring of bracts subtending several flowers or their supports, e.g. the heads of Compositae or the umbels of Umbelliferae.

involute – when the edges are rolled inwards spirally on each side.

keel – the lower united petals of a papilionaceous flower; a central dorsal ridge; hence *keeled* (of leaves) – ridged like the bottom of a boat.

labellum – a lip; referring to one of the petals which differs in size, shape or ornamentation from the other petals, e.g. the middle petal in Orchidaceae.

lacerate – torn; irregularly cleft or cut.

laciniate – deeply and irregularly divided into narrow pointed lobes.

lamina – the blade of a leaf or petal.

lanate – woolly, with long intertwined curly hairs; cf. lanuginose.

lanceolate – lance-shaped, rather narrow, tapering to both ends with the broadest part below the middle.

lanuginose – woolly or cottony; downy, the hairs somewhat shorter than in lanate.

lax – loose, the opposite of congested.

lateral – on or at the side.

lemma – in grasses, the flowering glume, the lower of the two bracts enclosing the flower.

lenticel – a lenticular corky spot on young bark, corresponding to an epidermal stoma; hence *lenticellate* – having lenticels.

lenticular – convex on both faces and lens-shaped in outline.

ligule – a strap-shaped organ or body, such as the limb of the ray florets in some Compositae, or the often membranous outgrowth at the inner junction of leaf-sheath and blade in grasses, etc., the latter sometimes represented by a fringe of hairs.

linear – long and narrow with ± parallel margins.

locule – compartment or cell of an ovary, anther or fruit.

loculicidal – splitting down the middle of each cell of the ovary.

lodicule – one of 2 or 3 minute scales appressed to the base of the ovary in most grasses; probably the vestiges of the perianth.

lyrate – pinnately lobed with the terminal lobe larger than the others.

membranous – thin, dry, flexible and translucent, not green.

mericarp – a 1-seeded portion split off from a syncarpous ovary at maturity, e.g. one of the two seed-like carpels of an umbelliferous fruit.

-merous – e.g. in 5-merous: having the parts in five.

midrib – the main rib or central vein which runs from the base to the apex of a leaf or leaf-like structure.

monoecious – having unisexual flowers with both sexes borne on the same plant.

mucro – a short sharp abrupt spur or spiny tip; hence *mucronate* – with a mucro; dimin. *mucronulate.*

muricate – rough with short hard points.

nectary – a nectar-secreting gland, often appearing as a protuberance, scale or pit.

node – that point on a stem which normally bears a leaf or leaves.

nut – a dry indehiscent one-seeded fruit usually with a hard woody or leathery wall.

obconical – conical but attached at the narrower end.

obcordate – inversely heart-shaped, with the notch at the apex.

oblanceolate – reversed lanceolate, i.e. lance-shaped with the broadest part above the middle.

oblong – longer than broad with the sides nearly or quite parallel for most of their length.

obovate – reversed ovate, the distal end the broader.

obovoid – egg-shaped but attached at the narrower end.

obtuse – terminating gradually in a ± rounded end, blunt.

operculate – with a cap or lid, e.g. in the circumcissile dehiscence of the fused perianth segments in *Richea.*

orbicular – rounded in outline with length and breadth about the same.

orifice – the mouth of a cavity, e.g. the opening at the top of the leaf sheath in grasses.

oval – broadly elliptical with the width greater than half the length.

ovary – that part of the pistil which contains the ovules, after fertilization forming the fruit containing the seeds.

ovoid – a solid resembling an egg, i.e. ovate in outline.

ovule – a structure containing the egg and developing into the seed after fertilization.

palea – the upper of the two bracts enclosing the grass flower.

pallid – pale.

palmate – lobed or divided in the manner of an outstretched hand, with the sinuses between the lobes pointing to the place of attachment.

palmatifid – cut about half way down in a palmate form.

palmatisect – cut in a palmate fashion nearly to the petiole.

panicle – strictly a compound or branched raceme but often applied to any branched inflorescence; hence *paniculate* – resembling a panicle.

pannose – having the appearance or texture of dense felt or woollen cloth of very close texture.

papillae – small nipple-shaped projections; adj. *papillose*.

pappus – tuft of hairs, bristles or scales as at the top of the achene of Compositae.

patent – spreading.

pedicel – the stalk of a single flower or spikelet; hence *pedicellate* – stalked.

peduncle – the stalk of an inflorescence or partial inflorescence; hence *pedunculate* – with a peduncle.

peltate – of a flat organ with its stalk attached on the undersurface, not on the margin.

percurrent – extending the whole length of an organ but not beyond it; cf. excurrent.

perennial – with a life span of more than two years.

perianth – the calyx and corolla together; used especially to describe structures which may be calyx or corolla or both.

perigynous – borne or arising from around the ovary and not beneath it, as when calyx, corolla and stamens arise from the edge of a cup-shaped hypanthium.

persistent – remaining attached, not falling off.

petal – a member of the inner series of perianth segments, if differing from the outer series, and especially if brightly coloured; hence *petaloid* – resembling a petal.

petiole – the stem of a leaf.

phyllary – an involucral bract in the Compositae.

pilose – with soft weak hairs, less dense than hirsute.

pinna (pl. *pinnae*) – a primary division or leaflet of a pinnate leaf.

pinnate – used in reference to a leaf composed of more than three leaflets arranged in two rows along a common stalk or rachis.

pinnatifid – pinnately lobed on both sides about half way to the midrib.

pinnatipartite – pinnately lobed on both sides more than half way to the midrib.

pinnatisect – pinnately lobed on both sides down to, or almost to, the midrib.

pinnule – a secondary pinna.

pistil – a unit of the gynoecium comprised of ovary, style (when present), and stigma; it may consist of 1 or more carpels.

plumose – resembling a feather or plume.

pod – a general term for a dehiscent dry fruit, e.g. the legume of Leguminosae or the siliqua of Cruciferae.

polygamous – having male, female and hermaphrodite flowers on the same or different plants.

polygamo-dioecious – polygamous but chiefly dioecious.

procumbent – lying loosely on the surface of the ground but not rooting.

produced – projecting.

proliferous – producing adventitious leafy shoots or buds capable of growing into new plants.

prophyll – a leaf-like or bract-like structure, often 2-keeled, found e.g. at the base of a branch.

prostrate – a general term for lying flat on the ground.

proximal – towards the attached, as opposed to the free or distal end of an organ.

pteridophyte – a vascular plant without flowers or seeds, reproducing by spores.

puberulent – minutely pubescent or downy, scarcely visible to the naked eye.

pubescent – covered with short soft hairs; downy.

punctate – dotted or shallowly pitted, often with glands; dimin. *puncticulate* – minutely dotted.

pungent – sharply and stiffly pointed, capable of pricking.

pyriform – pear-shaped.

quadrifarious – arranged in 4 close-set rows along an axis.

raceme – an indeterminate inflorescence consisting of a central rachis bearing a number of flowers with pedicels of nearly equal length; hence *racemose* – resembling a raceme or in racemes.

rachis – an axis bearing flowers or leaflets; dimin. *rachilla.*

radiate – spreading from or arranged around a common centre; having ligulate ray flowers as in Compositae.

radical (of leaves) – arising from the base of a stem or from a rhizome.

raphe – the united portions of the funicle and outer integument in an anatropous ovule.

ray – the marginal, as opposed to the disc florets in Compositae; one of the radiating branches of an umbel.

receptacle – the expanded portion of the axis which bears the floral organs; torus.

recurved – bent or curved downward or backward.

reflexed – abruptly bent downward or backward.

regular – radially symmetrical; actinomorphic.

reniform – kidney-shaped.

repand – weakly sinuate.

reticulate – marked with a network, usually of veins.

retrorse – directed backwards or downwards; opposite of antrorse.

retuse – notched slightly at a usually obtuse apex.

revolute – rolled backward; with margin rolled towards lower side.

rhachis – see rachis.

rhizome – an underground stem.

rhomboid – shaped approximately like the diamond in a pack of playing cards.

rootstock – subterranean stem or rhizome; often used when the rhizomes are relatively short and/or thick, cf. caulorhiza; or the butt or crown of the plant in the soil at the junction of root and stem.

rosette – a cluster of spreading or radiating basal leaves.

rostrum – a beak; hence *rostrate* – having a beak or beak-like projection.

rosulate – in rosettes.

rotate – applied to a gamopetalous corolla with a short tube and the lobes spreading at right angles to the axis like a wheel, e.g. corolla of *Myosotis.*

runcinate – saw-toothed or sharply incised, the teeth retrorse, i.e. pointing towards the base.

sagittate – enlarged at the base into two acute straight lines, like the barbed head of an arrow.

scaberulous – minutely scabrid.

scabrid – rough to the touch because of small harsh projections; dimin. *scaberulous* – minutely scabrid.

scale – any thin scarious body, e.g. a vestigial leaf or perianth segment.

scape – the flowering stem of a plant all the foliage leaves of which are radical; it may bear bracts or scales and may be one- or many-flowered; – adj. *scapigerous*.

scarious – thin dry and membranous, not green.

secund – said of parts or organs appearing to be arranged on one side of the axis.

sepal – one of the separate parts of a calyx, usually green and foliaceous; *sepaloid* –resembling a sepal.

septate – partitioned; divided by partitions.

septum (pl. *septa*) – a partition or cross-wall.

sericeous – silky with long soft straight closely appressed glossy hairs.

serrate – toothed like a saw; dimin. *serrulate* –finely saw-toothed.

sessile – without a stalk, e.g. a leaf without a petiole.

setaceous – bristle-like.

setose – bristly; with scattered ascending stiff hairs; dimin. *setulose* –with minute bristles.

sheath – any ± tubular structure surrounding an organ or part, e.g. the sheath of a grass leaf.

shrub – a low, usually several-stemmed woody plant; a bush.

sinuate – having a wavy margin in the plane of the blade.

simple – of one piece, not compound, e.g. a leaf not divided into leaflets or an inflorescence not branched.

sinus – the depression between two lobes or teeth.

solitary – single, only one from the same place.

sorus (pl. *sori*) – a cluster of sporangia on the leaves of ferns.

sparse – scattered.

spathulate – paddle-shaped or spoon-shaped in outline.

sphacelate – with brown or blackish speckling.

spike – an inflorescence consisting of a central rachis bearing sessile flowers or spikelets.

spikelet – a small spike; the unit of the inflorescence in grasses and sedges.

sporangium (pl. *sporangia*) – a structure containing spores.

sporophyll – a leaf or leaf-like organ, bearing sporangia.

spur – a hollow, usually ± conical projection from the base of a perianth segment, often a petal, or of a corolla.

stamen – the pollen-bearing organ, usually consisting of a stalk or filament and the anther containing the pollen; hence *staminate* –male.

standard – the large upper ± erect petal of a papilionaceous (pea-like) flower.

stellate – star-shaped or radiating like the points of a star.

stem – the main axis of a plant, leaf-bearing and flower-bearing.

stereome – in Compositae, the rigid central fibrous or herbaceous portion of the phyllary, extending upwards from the base.

stigma – the part of the pistil that receives the pollen, usually found at or near the tip of the style.

stipe – a short stalk; hence *stipitate* –borne on a short stalk.

stolon – a stem with elongate internodes that trails along the surface of the ground, often rooting at the nodes and usually capable of forming a new plant at its tip; hence *stoloniferous* –producing stolons.

stramineous – straw coloured.

striate – marked with fine grooves, ridges or lines of colour.

strigose – with stiff appressed straight hairs, pointing in one direction.

style – the part of the pistil between the ovary and the stigma.

stylopodium – a disc-like enlargement at the base of the style, as in some Umbelliferae.

subtend – to stand below and close to, e.g. as a bract below a flower.

subulate – awl-shaped; narrow, gradually tapering to a point and ± flattened.

superior (ovary) – with perianth inserted around the base, the ovary being free.

tepal – a segment or unit of those perianths not clearly differentiated into corolla and calyx.

terete – ± circular in cross section; narrowly cylindrical or tapering.

terminal – at the tip, apical or distal end.

ternate – arranged in or divided into three.

tetragonous – four-angled; of a solid body, four-sided in section with the angles rounded.

tomentose – with hairs compacted into a felty mass; hence *tomentum* – a covering of such hairs.

trifoliolate – having a leaf or leaves of 3 leaflets, as in most clovers.

trigonous – of a solid body, triangular in section, with the angles rounded.

triquetrous – of a solid body, triangular in section and acutely angled with the faces ± concave.

truncate – ending abruptly as though cut squarely across.

tuber – a swollen, usually subterranean part of a stem or root; hence *tuberous* – bearing or producing tubers.

tubercle – a ± spherical or ovoid swelling; hence *tuberculate* – furnished with tubercles.

turbinate – top-shaped.

turgid – swollen from fullness but not from air.

umbel – an inflorescence in which the pedicels all arise from the top of the main stem; also used of compound umbels in which the peduncles also arise from the same point; hence *umbelloid* – resembling an umbel.

umbonate – having a rounded projection or boss in the middle.

undulate – wavy in a plane at right angles to the surface.

utricle – a thin bladdery sac enveloping some fruits, e.g. in *Carex*, *Uncinia*, etc.

valvate – opening by valves or pertaining to valves; meeting by the edges without overlapping, as leaves or petals in the bud.

valves – the separate parts into which the wall of a fruit splits or divides to release the seeds.

velutinous – velvety; clothed with dense, erect, straight, moderately firm hairs.

versatile – attached near the middle and usually moving freely.

verticillate – arranged in whorls.

villous – with dense long soft hairs; shaggy.

whorl – more than two organs of the same kind arising at the same level, e.g. a whorl of leaves at a node.

wing – any membranous expansion attached to an organ; the lateral petals of a papilionaceous flower.

zygomorphic – having only one plane of symmetry; cf. actinomorphic.

Index

Numbers in bold type denote pages containing main references to families and genera and to descriptions of species; italicized epithets denote synonyms and numbers in square brackets denote plate numbers. Naturalized species are indicated by an asterisk.